Ptolemy's Universe

Urania   Astronomia   Ptolomeus

Ptolemy's Universe

*The Natural Philosophical and
Ethical Foundations
of Ptolemy's Astronomy*

Liba Chaia Taub

Chicago and LaSalle, Illinois

The frontispiece illustration is taken from: Sacro Bosco, *Textus De Sphera,* Paris, 1500. The cover illustration is from: [Sphaera mundi], [Venice], Erhard Ratdolt, 1485.

OPEN COURT and the above logo are registered in the U.S. Patent and Trademark Office.

© 1993 by Open Court Publishing Company

First printing 1993

All rights reserved. No part of this publication may be reproduced, stored in a retrieval system, or transmitted, in any form or by any means, electronic, mechanical, photocopying, recording, or otherwise, without the prior written permission of the publisher, Open Court Publishing Company, Peru, Illinois 61354.

Library of Congress Cataloging-in-Publication Data

Taub, Liba Chaia, 1954–
    Ptolemy's universe: the natural philosophical and ethical foundations of Ptolemy's astronomy / Liba Chaia Taub.
       p.   cm.
    Includes bibliographical references and index.
    ISBN 0-8126-9228-4. — ISBN 0-8126-9229-2 (pbk.)
    1. Astronomy, Ancient. 2. Cosmology. 3. Ptolemy, 2nd cent. I. Title.
QB16.T38   1993
520'.937'09015—dc20   92-40612
                                                                    CIP

To my parents

οἶδ' ὅτι θνατός ἐγὼ καὶ ἐφάμερος · ἀλλ' ὅταν ἄστρων
  μαστεύω πυκινὰς ἀμφιδρόμους ἕλικας
οὐκέτ' ἐπιψαύω γαίης ποσίν, ἀλλὰ παρ' αὐτῶι
  Ζανὶ θεοτρεφέος πίμπλαμαι ἀμβροσίης.

I know that I am mortal and living but a day.
  But, when I search for the numerous turning
    spirals of the stars
I no longer have my feet on the Earth,
  but am beside Zeus himself,
    filling myself with god-nurturing
    ambrosia.

> Anonymous epigram attributed to Ptolemy
> Epigram 9.577, from the Greek Anthology
> in *Further Greek Epigrams*

✳ Contents

✳ Acknowledgements — XI

✳ A Note on Editions and Abbreviations — XIII

✳ Illustrations — XV

✳ About the Illustrations — XVI

✳ Introduction — 1

1 ✳ Ptolemy and the Historians — 7

2 ✳ The Philosophical Preface to the *Syntaxis* — 19

3 ✳ The Hypotheses Underlying the *Syntaxis* — 39

> I. *That the heaven moves sphericalIy* — 45
> II. *That the Earth, taken as a whole, is sensibly spherical* — 61
> III. *That the Earth is in the middle of the heavens, with regard to the senses* — 71
> IV. *That the Earth has the ratio of a point relative to the size of the heavens* — 79
> V. *That the Earth makes no motion involving change of place* — 84
> VI. *That there are two different primary motions in the heavens* — 100

4 ✳ Ptolemy's Cosmology — 105

5 ✳ The Divinity of the Celestial Bodies and the Ethical Motivation for the Study of the Heavens — 135
> I. The Greek Cosmological Tradition and the Search for the Divine — 138
> II. The Ethical Motivation to Study Astronomy — 146

✳ Notes — 155

✳ Bibliography — 173

✳ Index — 185

Acknowledgements

It is a very pleasant duty to acknowledge those who have contributed to the completion of this book. The United Chapters of Phi Beta Kappa awarded me the Mary Isabel Sibley Fellowship (1988–89) which partially supported my work of revising and expanding my University of Oklahoma doctoral dissertation, which formed the basis of this book.

I am grateful to the many people who have helped me during the course of this project. Bernard Goldstein first pointed me in Ptolemy's direction. My professors at the University of Oklahoma (especially David Kitts, Steve Livesey, Mary Jo Nye, and Ken Taylor) and the University of Texas at Austin (particulary Alex Mourelatos and James Hitt), contributed significantly to my preparation in undertaking this work and carefully read and commented on earlier versions. Curtis Wilson, who was a reader on my dissertation committee, made many helpful suggestions. John

Dillon very kindly agreed to read page proofs and made a number of welcome comments and corrections. My colleagues at Loyola University of Chicago, Northwestern University, the University of Chicago, and the Adler Planetarium offered their support, encouragement, and suggestions. I am especially grateful to my mother and father for their patience and enthusiasm for this project, and to Marilyn Ogilvie and Alan Sable for their wisdom and sound advice. My husband, Niall Caldwell, contributed to the completion of this book in countless ways. In particular, I owe a great deal to Noel Swerdlow, who during the course of this work has read many versions, made helpful suggestions and corrections, and offered valued advice, encouragement, and friendship. I thank you all.

I am unable to personally thank three people, who I nevertheless must mention with gratitude. My grandmother, Minnie Feldman Flecker, introduced me to astronomy, by giving me a copy of Flammarion, and to history, through the constant example of her own reading. My teacher, Professor Aline Mackenzie Taylor (Tulane University) provided my first exposure to intellectual history, though I little realized it at the time. I fondly recall many happy hours of conversation with each of them. I did not know Professor Eric Aiton well, yet over a period of years he offered me much kind encouragement.

A Note on Editions and Abbreviations

Ptolemy's *Syntaxis* is cited according to the two-volume edition of Heiberg; his *Tetrabiblos* according to F. Boll and A. Boer. The *Planetary Hypotheses* has not survived in its entirety in Greek. Heiberg's edition in the *Opera Minora* includes the extant Greek text of Book One, a translation from the Arabic by L. Nix, and a translation of the Arabic text of Book Two by F. Buhl and P. Heegaard. The final portion of Book One, extant only in Arabic and Hebrew manuscripts, was omitted from the Heiberg edition. This second section of Book One is cited according to the edition and translation of the Arabic text by B. Goldstein. Other ancient authors are cited according to the editions named in the Bibliography.

Full details of modern works referred to will

be found in the Bibliography. They are cited in the text and notes by author's name and a short title.

These abbreviations have been used in citations of the following works:

CIAG	*Commentaria in Aristotelem Graeca.*
DK	H. Diels, *Die Fragmente der Vorsokratiker.*
DSB	*Dictionary of Scientific Biography.*
HAMA	Otto Neugebauer, *History of Ancient Mathematical Astronomy.*
OCD	*Oxford Classical Dictionary.*
RE	*Paulys Realencyclopädie der classischen Altertumswissenschaft.*

The standard text of the *Syntaxis,* and that cited here, is the critical edition of J. L. Heiberg. An earlier edition by N. Halma, *Composition mathématique de Claude Ptolémée,* also contains Halma's French translation and notes contributed by J. B. J. Delambre. A German translation was published by K. Manitius, *Ptolemäus, Handbuch der Astronomie,* with a foreword and corrections by O. Neugebauer. An English translation by R. Catesby Taliaferro, *The Almagest by Ptolemy,* is in the series *Great Books of the Western World.* The annotated English translation, *Ptolemy's Almagest,* by G. J. Toomer, has now become the standard. In addition to the edition of the *Tetrabiblos* by Boll and Boer, another edition, along with an English translation by F. E. Robbins, is available in the Loeb Classical Library series.

✳ Illustrations

Armillary sphere frontispiece

Figure
3.1   The rising and setting of stars   46
3.2   The circular motion of the stars, Sun and Moon   47
3.3   Stars that rise and set   48
3.4   Stars do not move in straight lines   50
3.5   Stars that rise and set   51
3.6   The circle is greater than any other plane figure   59
3.7   Eclipses occur simultaneously for all observers, but not at the same hour   61
3.8   The Earth is sensibly spherical   62
3.9   The Earth is spherical   64
3.10 The stars appear to be the same distance from Earth   73
3.11 If the Earth were not in the middle of the heavens . . .   73
3.12 The Earth appears to be in the middle of the heavens   75

About the Illustrations

Those sections of the *Syntaxis* which are dealt with in detail here were particularly well known during the Middle Ages and the Renaissance. Indeed, they were far better known then than now, for commentaries on the physical hypotheses occupied an important place in elementary treatises on spherical astronomy. While Ptolemy provided no illustrations for these sections, medieval and Renaissance works often contained diagrams to supplement the discussion. I have chosen some rather charming woodcuts from fifteenth and sixteenth century editions of such works which nicely illustrate points made by Ptolemy and discussed further by his commentators. The illustrations are reproduced courtesy of the Adler Planetarium, History of Astronomy Collections.

Introduction

Claudius Ptolemy is most renowned for his contribution to mathematical astronomy, an achievement that ensured his place as one of the outstanding figures in the history of science. Appropriately, his work and what little can be learned concerning his life have been subjects of extensive study for many centuries; understandably, historians of science have emphasized the mathematical aspects of Ptolemy's work. However, within his astronomical writings Ptolemy presented some of his views on natural philosophy, epistemology, and ethics, even explaining the pertinence of his philosophical principles to his astronomy.

This study is concerned with the philosophical foundations, especially the natural philosophical and ethical aspects, of Ptolemy's astronomy. As such, it is a consideration of only part of his work and is not intended to cover material which has already been dealt with elsewhere. For a

2
PTOLEMY'S UNIVERSE

complete picture of Ptolemy's astronomy, the reader will find that the astronomical aspects have been discussed in detail by Pedersen in his *Survey of the Almagest;* many points have received further clarification in the explanatory material in Toomer's translation, *Ptolemy's Almagest.* In addition, the reader should consult Neugebauer's *History of Ancient Mathematical Astronomy,* as well as *Measuring the Universe* by van Helden.

The aim of the present work is the examination and understanding of the philosophical ideas which underlie Ptolemy's astronomical writings, with a view to determining his place within the Greek philosophical and scientific traditions.

Chapter one, "Ptolemy and the Historians," reviews earlier treatments of Ptolemy's intellectual orientation, which tended to characterize his point of view as "Aristotelian" or "Stoic." The phrase "the Aristotelian-Ptolemaic universe" occurs quite often in the literature of the history of science. The precise, intended meaning of this phrase is not clear; furthermore, it seems to suggest that Aristotle and Ptolemy shared a similar conception of the universe. A secondary goal of this work is the examination of the extent to which Aristotle's and Ptolemy's ideas about philosophy and physics coincide. The first chapter outlines the historiographical problems involved in the study of Hellenistic philosophy and the difficulties encountered in placing Ptolemy within any particular school of thought. The sources for the study of Hellenistic philosophy are regrettably fragmentary; very few texts survive in their entirety. Because of the paucity of surviving texts and biographical evidence, the establishment of Ptolemy's sources would be an unrealistic goal.

Rather, our understanding of Ptolemy's philosophical ideas must be primarily based on a close study of his own writings.

The Mathematical Treatise (ἡ μαθηματικὴ σύνταξις) is Ptolemy's most important astronomical work. While the Arabic-Latin name "*Almagest*" is widely used by later writers, I will refer to it here as the "*Syntaxis*," to emphasize my concern with Ptolemy's place within the Greek tradition. Chapters two and three are close examinations of the text of the *Syntaxis* itself. Ptolemy began the work with a philosophically oriented preface, in which he discussed the organization of knowledge, concentrating particularly on the nature of and relationships between physics, mathematics, and theology. Here, in the preface, we see the first indication that Ptolemy broke with Aristotle, for he argued that mathematics is the highest form of philosophy, rather than theology. Ptolemy also explained his ethical motivation for studying mathematics in general and astronomy in particular. (This motivation will be discussed in detail in the final chapter of this work.) For Ptolemy, the consideration of the nature of astronomy was not only important, but even prerequisite, to the study of astronomy itself. Chapter two focuses on the philosophical foundation provided by Ptolemy as a preface to his astronomical work.

Following the preface, Ptolemy outlined six physical hypotheses, or assumptions, which are essential to his astronomical models; these are discussed in detail individually in chapter three. The hypotheses constitute the basis of Ptolemy's cosmology, being concerned generally with the heavenly motions and the place, size, and shape of the Earth relative to the universe as a whole.

✦4 PTOLEMY'S UNIVERSE

While his ideas are recognizably part of the tradition of Greek natural philosophy, Ptolemy's statement of his assumptions is in many ways original. For example, his preoccupation with mathematical considerations is always evident in his discussion of these physical assumptions; this preoccupation distinguishes his treatment from that of the natural philosophers, including Aristotle.

As a mathematician, Ptolemy was not concerned with presenting a fully developed physical theory. He did, however, add further details and some refinement to his natural philosophy in his subsequent astronomical writings; chapter four examines these later works. The *Planetary Hypotheses* may be regarded as a companion piece to the *Syntaxis;* here, Ptolemy made it clear that his mathematical models are intended to have physical reality. In the *Planetary Hypotheses* he discussed in detail such physical questions as the order and distances of the planets, as well as the causes of their motions. It should be noted that Book Two of the *Planetary Hypotheses* is a direct attack against Aristotle's celestial physics. But perhaps most striking is Ptolemy's blurring of the Aristotelian distinction between the terrestrial and celestial regions, which occurs in several of his works. (Some of this blurring was accomplished by Aristotle himself; in the *Meteorology,* he suggested that celestial motions can affect the terrestrial realm.) The idea of an important correlation between terrestrial and celestial phenomena lies at the heart of the *Phases of the Fixed Stars,* the *Harmonics,* and the *Tetrabiblos.* The *Phases* and the *Tetrabiblos* are based on the notion that terrestrial events may be predicted from celestial motions, while the *Harmonics* describes the analogies between musical structures, human character, and the celestial bodies.

That the celestial bodies have special significance for the human soul is an idea which Ptolemy introduced in the philosophical preface to the *Syntaxis*. For Ptolemy, the celestial bodies were divine. He suggested that the study of their motions is an ethical endeavor, because by studying and emulating the celestial motions astronomy enables man to become as similar to the divine as is humanly possible. Ptolemy regarded mathematics as the highest form of philosophy; astronomy, that branch of mathematics which studies the celestial motions, he regarded as an ethical endeavor. For Ptolemy, astronomy teaches man to regulate his life in imitation of the order, constancy, and tranquility of the celestial motions.

Ptolemy shared his goal of becoming more divine, through the attainment of inner peace and tranquility, with many Hellenistic philosophers. His way of achieving this goal was somewhat unusual. However, his emphasis on the study of mathematics as an ethical endeavor was not unique in the history of Greek philosophy, for mathematics in general, and astronomy in particular, held a special place in Plato's ethical philosophy. Chapter five traces these themes in the history of Greek philosophy.

Geoffrey Lloyd concluded his recent book, *The Revolutions of Wisdom,* with the following observation (p. 336):

> the fact that in its beginnings, science was often explicitly concerned, if sometimes rath-

er naively, with the moral dimension of the activity of science itself reminds us, if we need reminding, that it originated in no merely intellectualist debate. Indeed, its offering an alternative world view, in the widest sense an alternative morality, was central to some of its confrontations with traditional wisdom.

Ptolemy's Universe is an attempt to elucidate the world view of Ptolemy in two senses; his physical cosmology is examined in detail, as well as the moral dimension of his astronomical work. Ultimately, the physical structure of Ptolemy's universe allows individual men to achieve moral progress. The analogies and influences between the celestial and terrestrial regions, between the divine and the human, may be studied and imitated by man. In doing so, man achieves some measure of divinity.

Ptolemy's conception of the celestial bodies as divine beings shaped his ideas about natural philosophy, epistemology, and ethics; these ideas in turn play a significant role in his study of the heavens. By understanding Ptolemy's philosophy, we enlarge our appreciation of his astronomical achievement; by examining the place of Ptolemy's ideas within the broader context of Greek philosophy, mathematics, and culture, we gain insight into the nature of Greek scientific thought.

✳ 1
Ptolemy and the Historians

Little is known about the life of Ptolemy. As O. Neugebauer ironically remarked, we are in "the fortunate situation that almost the only source for a biography of Ptolemy is Ptolemy's work itself." The brief mention of Ptolemy given by the sixth-century commentator, Olympiodorus, provides no help in establishing dates for Ptolemy's life. The tenth-century historical lexicon known as the *Suda,* quoting Hesychius of Miletus (sixth century A.D.), states that Ptolemy lived during the time of the king Marcus, presumably Marcus Aurelius, emperor from A.D. 161–80. Other sources of information on Ptolemy's life are Byzantine, Arabic, or by anonymous scholiasts, whose reliability and information cannot be assessed since there are no early sources against which they can be checked. Faced with this paltry evidence, historians have deduced approximate dates for Ptolemy's life from observations reported in his own work.[1] For our purposes, it is sufficient to know that the evidence suggests that Ptolemy lived during the second century A.D.

The *Suda* describes Ptolemy as an Alexandrian philosopher. Alexandria was the only place mentioned by Ptolemy as a site from which he made observations and was the place he used to establish the times of the positions of the celestial bodies.[2]

Of Ptolemy's education, nothing is known. While there has been some speculation that the Theon mentioned by Ptolemy as having given him observational information may have been his teacher, there is nothing to support this interpretation of their relationship. The lack of information about Ptolemy's education is particularly unfortunate for, as John Dillon has noted, on

> the general question of the influence of earlier authorities on later ones, there is one fact of life that is liable to be overlooked by modern scholars. That is that, in this period —as indeed in most periods of intellectual development until quite modern times— one is influenced primarily by the doctrine of one's own teacher, and one sees the development of philosophy up to one's own time through his eyes. One may indeed read the original texts, but one reads them initially under the guidance of one's teacher, who read them under *his* teacher, and so on.[3]

Without knowledge of the identity and philosophical views of Ptolemy's teachers, recourse must be made to the examination of his own writings.

Attention has been understandably concentrated on Ptolemy's astronomical achievements and on his major work, the mathematical treatise referred to by Ptolemy himself as *The Mathemati-*

cal Syntaxis (ἡ μαθηματικὴ σύνταξις). In 1894, Franz Boll remarked on the general absence of any work evaluating Ptolemy's position in the history of Greek philosophy.[4] While Boll alleviated this circumstance somewhat, he limited his attention chiefly to the *Tetrabiblos* and Ptolemy's astrology.

Because of the overriding interest in Ptolemy's mathematical work, now, nearly a century later, little has been added to our knowledge of Ptolemy's relation to Greek philosophy. One student of Ptolemy's philosophical interests was Friedrich Lammert. However, Lammert was primarily interested in *On the Criteria* (a work whose attribution to Ptolemy has been questioned) and Ptolemy's relation to Stoic philosophy.[5]

In general, historians of science have tended to characterize Ptolemy's philosophical and cosmological views as "Aristotelian," and one could cite many examples of this assumption. However, Neugebauer recognized that Ptolemy's philosophical ideas may not have stemmed wholly from Aristotle, and briefly stated that it "is not surprising that Ptolemy followed peripatetic or stoic doctrines but with a certain eclectic attitude." Without discussing Ptolemy's views, he concluded that "these philosophical theories are without importance for his actual astronomical work."[6] Neugebauer is probably correct in his assessment of Ptolemy's philosophical views as eclectic. However, the labels "Aristotelian," "peripatetic," and "stoic," are simply not sufficiently descriptive to provide an understanding of Ptolemy's philosophical and physical statements. They are, at best, vague and ambiguous terms.

In order to assess the validity of attaching such

10
PTOLEMY'S UNIVERSE

labels to Ptolemy, these terms must first be examined and defined within a specific historical context, that of second-century Greek philosophy. Unfortunately, our knowledge of the period is fragmentary. Furthermore, there has been a tendency on the part of historians of philosophy to jump from the excitement of the fourth century and the early Hellenistic philosophical schools to the glamour of Neoplatonism, without much enthusiasm for the centuries which lie between. It is difficult to know how to place the philosophical views of a particular individual, in this case Ptolemy, when for the period itself there is a general problem in defining philosophical orientations. While a detailed investigation of second-century Greek philosophy is not in order here, nevertheless, some difficulties with the terminology must be addressed.

In many accounts of the philosophical schools of this period, the reader is haunted by the occurrence of the word "eclectic." "Eclectic" is a seemingly apt term to describe the climate of second-century Greek philosophy, but this label is not without ambiguities. Edward Zeller, who did recognize the existence of various schools of philosophy, nevertheless characterized Greek philosophy of the four centuries from the second century B.C. to the second century A.D. as eclectic. He ascribed this general development in Greek philosophy both to external causes, which included the increased contact between Roman intellectuals and Greek philosophers, and to an intellectual reorientation internal to philosophy itself, particularly an increased interest in religion. This philosophical interest in religion was paralleled by, and probably part of, a general increase in religious activity during the same period, exemplified by the rise of various eastern religions throughout Rome.[7]

When "eclecticism" is used to describe the general tenor of Greek philosophy during the period, it signifies "a type of approach to philosophy which consists in the selection and amalgamation of elements from different systems of thought."[8] The term also is applied both to a particular school[9] and to a large group of individuals who were not actually part of that school, but who selectively chose philosophical ideas from a variety of sources to incorporate into their own writings. Among the ancient writers called "eclectic" in this second sense are Panaetius, Antiochus of Ascalon, Cicero, Posidonius, Plutarch, and Seneca.

Since the eclectic philosopher by definition chooses his philosophical ideas from a variety of sources, some of the major sources of ideas available in the marketplace of the period must be mentioned. The major philosophical schools of the second century A.D. included the Academy, the Peripatos, and the Stoa. Skepticism and Neopythagoreanism were two important philosophical movements which influenced various and diverse philosophers, not just the members of a particular school. Furthermore, some important and influential figures during the period seem not to have been associated with any specific school and defy easy categorization, for example, Philo of Alexandria.

The Academy was founded by Plato and the Peripatos by Theophrastus, both schools having been established in the fourth century B.C. The school known as the Stoa was started by Zeno of Citium in the third century B.C. Each of these schools experienced changes and underwent differing degrees of development in the centuries following the deaths of their founders. The major schools of philosophy during the Hellenistic period did not remain the same as they had been at the time of their founding. For example, in the third century A.D. Diogenes Laertius recognized three different "Academies": the "Old" Academy founded by Plato, the "Middle" Academy founded by Arcesilaus of Pitane (third century B.C.), and the "New" Academy" founded by Lacydes of Cyrene (third century B.C.). Modern historians have adopted their own terminology to distinguish the Platonism associated with the "Old" Academy from "Middle" Platonism and the "Old" Stoa from the "Middle" Stoa.[10]

The term "skepticism," like "eclecticism," describes both a general trend in Greek philosophy and a specific philosophical movement which lasted about five hundred years, from Pyrrho of Elis (fourth century B.C.) and Arcesilaus of Pitane to Sextus Empiricus (late second century A.D.), the aim of which was to avoid philosophical dogmatism. The Skeptics also produced a collection of arguments against the various dogmatic positions of the different Greek philosophical schools and against the possibility of knowledge itself.[11] The revival of Pythagoreanism in Rome and Alexandria during the first century B.C., both as a philosophical school and a religion, combined religious interests with elements of Academic, Peripatetic, and Stoic philosophy.[12]

In the second century A.D. there was a blurring of divisions between the different philosophical schools. In many cases there was a shared vocabulary, shared interests, and shared sources and influences,[13] and it is therefore difficult to point to the influence of one or another particular philosopher, with a credible degree of certainty. The stamp of Aristotle has been noticed, to varying degrees, in all of these schools, but his influence was by no means predominant. Indeed, certain elements which might well have been labeled "Aristotelian" in an earlier period were, by the second century A.D., part of the shared intellectual heritage of Greek philosophy and should be regarded as "Greek" (with the understanding that this influence stretched to some Romans as well, whether they were writing in Latin or in Greek), rather than as "Aristotelian."

The philosophy of this period may be more fruitfully characterized as "syncretic" rather than as "eclectic." Some of the motivation for adopting this term lies in its application to the description of religious experience in the Greco-Roman world during this period; by adopting this term (which is often used to describe the Hellenistic approach to religion), the similarities between these two movements are emphasized. Furthermore, the emphasis placed by philosophers on theological and ethical questions is suggested.[14] The word "eclectic" suggests that personal taste and expediency are the operative values; the term "syncretic" more justly describes the efforts towards the reconciliation and unification of different philosophical views which occurred during the period.

Ptolemy himself was rather reticent about mentioning those thinkers who influenced his own ideas. It is striking that in the entire *Syntaxis*, Ptolemy mentioned only one philosopher by name: Aristotle. The reference to Aristotle is a passing one, occurring at the beginning of Book One. Ptolemy did not mention any of Aristotle's works by name. Which specific works of Aristotle, if any, Ptolemy was familiar with is not clear. The extent of the influence of Aristotle and his writings on Ptolemy is a question still to be considered.

Since the question of whether Ptolemy's viewpoint was Aristotelian is at issue, the meaning of the term "Aristotelian" must be addressed. Many things may be described by the term "Aristotelian." In this work, Aristotelian refers to the point of view of Aristotle as represented by the extant corpus. Of course, the extent of the influence of Plato on Aristotle, that is, the extent to which Aristotle was a Platonist, cannot be overlooked but should not sidetrack us here.[15] In order to determine the extent of Ptolemy's Aristotelianism, the question must be asked: what writings of Aristotle were available in the second century?

Andronicus of Rhodes (first century B.C.) edited those writings of Aristotle which have survived as the Aristotelian corpus. The edition by Andronicus represents only a part of the entire body of writing considered to have been written by Aristotle, as is clear from the ancient lists of Aristotle's writings. And these would not have been the only writings in circulation which were attributed to Aristotle. In addition to whatever Aristotelian writings Ptolemy may have had access to that are no longer extant, various ancient commentaries on the writings of Aristotle would have been available.[16]

To have knowledge, for example, of Aristotle's division of philosophy, which Ptolemy demonstrated in chapter one of the *Syntaxis,* does not necessarily mean that Ptolemy read the *Metaphysics.* He might have learned of Aristotle's ideas from the works of other writers who were influenced by Aristotle, including those in the epitome and commentary tradition, such as Eudorus of Alexandria (first century B.C.), who wrote commentaries on the *Categories* and the *Metaphysics,* and Aspasius (second century A.D.), who wrote commentaries on several of Aristotle's writings, including the *Metaphysics, Physics,* and *On the Heavens.* These commentaries do not survive.[17]

Likewise, Ptolemy may have read Plato's *Timaeus.* Or, once again, he may simply have read another author who reported or incorporated the ideas contained in that work, for example, the *Platonikos* of Eratosthenes of Cyrene (third century B.C.), which seems to have been a commentary on the *Timaeus,* or the commentaries on the *Timaeus* by Eudorus of Alexandria (now lost), or Atticus (second century A.D.), or Theon of Smyrna (second century A.D.), or a commentary attributed to Albinus (second century A.D.).[18] In any event, Ptolemy did not mention Plato, or any of his works, by name.

Signs of the influence of Stoic philosophy on Ptolemy have been detected by several scholars. Lammert discussed the influence of the philosophers of the Middle Stoa on Ptolemy's views as expressed in *On the Criteria.* Boll and Pedersen have both specifically pointed to the influence of the Stoic Posidonius in a work written after the *Syntaxis,* the *Tetrabiblos.*[19]

Posidonius of Apamee was the student of Panaetius and the teacher of Cicero. He wrote works on many topics, including logic, ethics, physics, history, and geography. Some of the reported titles of his works include *Physical Discourse, On the Cosmos, Meteorology, On the Size of the Sun, On Gods.*[20]

None of the works of Posidonius are extant, which makes the task of determining the true extent of his influence on Ptolemy difficult, if not impossible. There is even difficulty ascertaining which fragments are rightly attributed to Posidonius. As Edelstein remarked, the "disagreement concerning those documents which are supposed to echo the philosophy of Posidonius is so great that it would be quite impossible to assume the Posidonian origin of any passage without incurring objection." The textual problems have led to controversies regarding the reconstruction of his philosophical ideas and his influence on other writers. In any case, the possibly important influence of Posidonius on Ptolemy would not necessarily qualify for the label "Stoic." Rist has hinted that Posidonius frequently "misunderstood" the original Stoic doctrine. Edelstein went so far, when speaking of Posidonius's physical system, as to remark that "compared with the general Stoic system it is heretical." He pointed to the influence of both Plato and Aristotle on Posidonius, but stated that although Posidonius's physical system was "influenced by Platonic and Aristotelian thought, it is neither Platonism nor Aristotelianism; it is original."[21]

Ultimately, the investigation and tracing of Ptolemy's ideas to earlier philosophical sources is not the aim here. Given the lack of evidence and surviving writings this would be an unrealistic goal. Rather, the achievement of a better understanding of Ptolemy's philosophical and physical ideas, as expressed in his astronomical writings, is the aim of this work, with the secondary aim being an examination of the degree of congruence to Aristotle's ideas.

How valid is the application of the label "Aristotelian" which has clung to Ptolemy for so long? It is not sufficient to say that because most second-century philosophers incorporated "Aristotelian" elements, as well as ideas and approaches from other philosophers and schools, into their thinking, this label has no particular meaning for Ptolemy. Instead, the texts themselves must be examined to see whether, in fact, there is a preponderance of ideas and methods which could be called "Aristotelian" at all. Labels are used in what follows, but in each case an effort is made to tie the use of the label to an example taken from a particular author and text. The label "Aristotelian" used here indicates that there exists a clear connection to a particular idea attributed to Aristotle, known either through his extant writings or through commentators. Similarly, the term "Stoic" must be linked to a particular philosopher or text. What specific ideas of Ptolemy were developed under the influence of Aristotle? In what ways did Ptolemy deviate from "Aristotelian" philosophy? Which Stoic philosophers were important in shaping his ideas? These are, at this point, unanswered questions. In order to answer these questions and assess the influence of Aristotle and other philosophers in the writings of Ptolemy, his familiarity with the philosophical literature must first be considered. Any discussion of philosophical content and various influences present in Ptolemy's writings must therefore rely on an examination and interpretation of the texts themselves.

✻ 2
The Philosophical Preface to the *Syntaxis*

Chapter one of the first book of the *Syntaxis* is a compact and philosophical preface. It is here that Ptolemy presented his most thorough discussion of epistemology and methodology. Ptolemy wove together two themes, both stylistically and conceptually, throughout this preface: the organization of philosophy and his own motivation for studying mathematics. Ultimately, he explained that mathematics was a very special type of philosophy.

Ptolemy began his discussion of philosophy by noting that true philosophers have rightly distinguished practical from theoretical philosophy, noting that theoretical philosophy requires education, while practical philosophy does not. In pointing to what he called a great difference between practical and theoretical philosophy, Ptolemy was making a distinction similar to that drawn by Aristotle in the *Nicomachean Ethics*. Here, Aristotle stated that intellectual virtue has its birth and growth from teaching and therefore requires experience and time, but that ethical virtue comes from custom or habit. But, Aristotle was describing the different ways in which a person acquires virtue; he was not discussing philosophy. He explained that some virtues are intellectual and others are ethical and gave examples of each. Wisdom, understanding, and prudence are all intellectual virtues; generosity and moderation are the two examples he gave of ethical virtues.[1] Ptolemy seems to have equated practical and theoretical philosophy with ethical and intellectual virtue.

Wisdom ($\sigma o \phi i \alpha$), one of the intellectual virtues named by Aristotle, was for him a particular kind of knowledge ($\dot{\epsilon} \pi \iota \sigma \tau \acute{\eta} \mu \eta$). In the *Metaphysics*, Aristotle divided knowledge into different types, the practical, the productive, and the theoretical.[2] Ptolemy, on the other hand, divided philosophy into only two types, the practical and the theoretical. If Ptolemy's bipartite division of philosophy was based on a division previously made by Aristotle, it is striking that Ptolemy did not include the third Aristotelian type of knowledge, the productive, in his own discussion.

Did Ptolemy, in his discussion of practical and theoretical philosophy, have in mind the sort of distinction that Aristotle had made between practical and theoretical knowledge? In the passage from the *Metaphysics*, there seems to be some degree of synonymy between $\dot{\epsilon} \pi \iota \sigma \tau \acute{\eta} \mu \eta$ and $\phi \iota \lambda o \sigma o \phi \iota \alpha$ (philosophy). At 1026a27–31 Aristotle explained that if there exists no substance other than those which are composed by nature, then physics would be the first kind of knowledge. But, if there exists some immovable substance, this would be prior, and the study of it would be first philosophy. Because of the parallel construction here, ἐπιστήμη and φιλοσοφία do look synonymous.

But even if a reading of this passage could lead to the idea that there was an equation between ἐπιστήμη and φιλοσοφία made by Aristotle, Ptolemy's conception of practical philosophy seems to have been quite different from Aristotle's notion of practical knowledge. Ptolemy stated that practical philosophy did not require education, while in the *Nicomachean Ethics* Aristotle made it quite clear that the study of politics (a type of practical knowledge) was not only necessary but very demanding.[3]

Ptolemy credits Aristotle with the division of theoretical philosophy into physics, mathematics, and theology. Regarding the three-way division of theoretical philosophy, Boll believed that Ptolemy, when he wrote his preface, undoubtedly had before him Aristotle's *Metaphysics* Book Six, the passage in which Aristotle presented this tripartite division. It is worth looking in detail at that passage from the *Metaphysics:*

> That physics, then, is theoretical, is apparent from these considerations. Mathematics also is theoretical; but whether its objects are immovable and separable [from matter], is not at present clear. It is clear, however, that it studies some mathematical objects as immovable and as separable. But if there is something which is eternal and immovable and separable, it is clear that the knowledge of it belongs to a theoretical science, but not, however, to physics (for physics is concerned with certain movable things) nor to mathe-

matics, but to a branch of *theoretike* prior to both. For physics is concerned with things which are separable but not immovable, and some parts of mathematics deal with things which are immovable, but probably not separable, but embodied in matter. But the first branch of *theoretike* is concerned with things which are both separable and immovable. It is necessary that all causes be eternal, but especially these; for they are the causes of what is visible of the gods. Thus there must be three theoretical philosophies: mathematics, physics, and theology, for it is clear that, if the divine exists anywhere, it exists in things of this kind.

Aristotle defined what he called the first branch of theoretical philosophy as dealing with things which are eternal (ἀίδιον), separable (χωριστόν), and changeless (ἀκίνητον). He called this "first" branch theology. For Ptolemy, theology was that branch of theoretical philosophy which is concerned with investigating the first cause of the first motion of the universe (τὸ τῆς τῶν ὅλων πρώτης κινήσεως πρῶτον αἴτιον), which may be regarded as an invisible and immutable divinity (θεὸν ἀόρατον καὶ ἀκίνητον ἂν ἡγήσαιτο). He stated that this kind of activity located "upwards, somewhere around the highest things of the cosmos, may only be imagined and is absolutely separated from sensible reality." Aristotle and Ptolemy envisioned the divine located in the same region; in the *Physics*, Aristotle stated that the Unmoved Mover occupies the circumference. But, while Ptolemy and Aristotle apparently had the same thing in mind, Ptolemy's language is not particularly imitative of Aristotle's.[4]

Similarly, Ptolemy's definition of physics does not suggest that he was using this passage from the *Metaphysics* as a guide. According to Ptolemy, physics examines the form or nature ($εἶδος$) of the material and always-changing quality ($τῆς ὑλικῆς καὶ αἰεὶ κινουμένης ποιότητος$). He gave examples of what sorts of things these qualities are: such things as white, hot, sweet, and soft.

In the passage from the *Metaphysics* quoted above, Aristotle stated that physics is concerned with what is separable but not changeless ($χωριστὰ μὲν ἀλλ' οὐκ ἀκίνητα$); he did not discuss the qualities of matter. That change ($κίνησις$) is the key concern of physics is also shown in another passage from the *Metaphysics*: "to physics one would assign the study of things not as being, but rather as sharing in change."[5] Ptolemy's discussion of what constitutes physics emphasized the qualities, while Aristotle emphasized change. Yet, Ptolemy agreed to some extent that change is characteristic of the subject matter of physics, for he described those qualities which physics studies as "always changing." Once again, Ptolemy mentioned the location of the subject matter of physics; he stated that it is "for the most part among those things which are corruptible and below the sphere of the Moon."

Ptolemy deviated even more strikingly from Aristotle in his definition of mathematics. According to Ptolemy's definition, mathematics clarifies quality with respect to form and motion from place to place (i.e., the subject matter of physics) by investigating shape, quantity, and size, as well as such things as place and time. This definition is quite clear and concise, while Aristotle admitted to some uncertainty concerning just what the objects of mathematics were. Aristotle explained that at present it is not clear if the objects of mathematics are changeless and separable, but that it is clear that mathematics studies some kinds of mathematical things as changeless and separable.[6]

In the *Physics,* Aristotle described the study of the mathematician as being concerned with surfaces and solids, points and lines, mentally separated from physical bodies. Another passage from the *Metaphysics* is relevant here, in which Aristotle explained that the mathematician makes a study of abstract things, eliminating all sensible things. In a further passage, he stated that mathematics is concerned with that which remains the same, but is not separable.[7] The concise definition of mathematics given by Ptolemy in the preface contrasts with Aristotle's rather tentative descriptions of mathematics in the *Metaphysics.* Ptolemy seems to have been very certain as to what mathematics was for him, while Aristotle, in the *Metaphysics* at least, was more cautious, and perhaps more thoughtful, about offering pronouncements as to the nature of mathematics.

Ptolemy's definitions of physics, mathematics, and metaphysics do not share much with the passage in Aristotle's *Metaphysics* which Boll pointed to as having been so important to Ptolemy in his writing of the preface.[8] But while these definitions were apparently not composed as Ptolemy stood over a copy of the *Metaphysics,* there is something about them which has a familiar ring, which suggests, broadly speaking, some "Aristotelian" influence. These definitions may represent Ptolemy's interpretation of Aristotle's division of philosophy, based on his own reading. It is interesting that Ptolemy named Aristotle as the author of this tripartite division of theoretical philosophy. For whatever reason, Ptolemy was desirous of having an Aristotelian stamp to this preface; he did, however, add some non-Aristotelian emphases and twists.

The sharpest twist occurs in the passage which follows, where Ptolemy, in effect, stood Aristotle's organization of knowledge on its head. After defining each branch of theoretical philosophy, Ptolemy described the relationships among theology, physics, mathematics as follows. Whereas theology deals with what can only be thought about and physics is concerned with those things among perishable bodies and below the lunar sphere, the subject matter of mathematics falls in between the other two branches of theoretical philosophy. This is so for two reasons.

First, the subject matter of mathematics can be thought of both with and without the senses; presumably, what Ptolemy meant here is that mathematics can be studied with such visual aids as diagrams, but need not rely on the senses. Second, and more importantly, the subject matter of mathematics is fundamentally linked to the subject matter of both physics and theology, for it is an attribute of all things, both those which are mortal and those which are immortal. In mortal things, which are always changing according to their inseparable form, it changes together with them; in immortal things it preserves the unchanging form as unchanged. Thus, what is mathematical serves an especially important function with regard to the divine realm; Ptolemy stated that it is the essence ($o\dot{v}\sigma\acute{\iota}\alpha$) of the type of inquiry known as mathematics which preserves ($\sigma v v \tau \eta \rho o \tilde{v} \sigma a v$) the unchanging form of those eternal things which have an aethereal nature. The $o\dot{v}\sigma\acute{\iota}\alpha$ of mathematics is the subject matter of mathematics, not the discipline of mathematics itself, but the essence or immutable reality which mathematics investigates.

As one might expect, Ptolemy accorded the study of mathematics the highest status of all of the three branches of theoretical philosophy. This was certainly not an Aristotelian point of view; Aristotle stated that theology is the best ($βέλτιστον$), or highest, of the branches of theoretical knowledge, because it is concerned with the most honored of existing things, and each kind of knowledge is called better or worse according to its proper object of study.[9] For Aristotle then, the study of divine things, theology, was the highest and best form of theoretical philosophy.

Ptolemy made his statement regarding the primacy of mathematics rather self-consciously, for he then presented arguments to support his claim. He made the rather radical, and certainly non-Aristotelian, statement that both theology and physics should be called conjecture ($εἰκασία$) rather than knowledge ($κατάληψις\ ἐπιστημονική$). He reasoned that theology should not be called knowledge because of the invisibility and ungraspability ($διὰ τὸ παντελῶς ἀφανὲς αὐτοῦ καὶ ἀνεπίληπτον$) of its subject; physics should not, because of the instability and lack of clarity of matter ($διὰ τὸ τῆς ὕλης ἄστατον καὶ ἄδηλον$). The reader is left with the clear impression that, so far as Ptolemy was concerned, mathematics represents the only true kind of knowledge.

For Ptolemy, mathematics can, and must, work together with the other branches of theoretical philosophy. Earlier he had stated that what is mathematical preserves the divine nature which is the subject matter of theology. Thus, the unchanging form of those things which are eternal and have an aethereal nature, those things which are the subject matter of theology, relies on the οὐσία (essence) of mathematics for its preservation. In order to understand what is eternal and divine, one must also understand that which keeps the divine eternal and unchanging. Therefore, there is an intimate and necessary connection between theology and mathematics. Furthermore, the way in which mathematics can work with the study of physics is also not a matter of chance.

Ptolemy began his explanation of how mathematics can make a significant contribution to physics with the statement that local motion is the key to the acquisition of understanding concerning material nature. In the *Physics,* Aristotle had distinguished three kinds of motion or change: change of quality, change of quantity, and change of place. He stated that where there is change of quality or change of quantity, there is necessarily change of place, since change of place is the primary kind of motion, taking precedence over the others. Aristotle acknowledged that the idea of the primacy of local motion had been around for a long time; he noted that many of his predecessors who had studied motion had treated local motion as the primary principle of change.[10]

By stating that every individual thing about material nature becomes apparent from the idiosyncracies of its change of place, Ptolemy was well within the Greek philosophical tradition concerning the treatment of motion, which Aristotle subscribed to as well. Ptolemy went on to describe changes of place in geometrical terms. What is corruptible can be distinguished from what is incorruptible by its motion; things either move in a straight line or in a circle. Similarly, what is heavy and light, passive and active, can be distinguished by whether its motion is toward or away from the center.[11]

Of course, Ptolemy used the Aristotelian explanation of motion as a background and, in that sense, his statement appears to be an extrapolation from Aristotle's *Physics,* although no one particular passage from that work seems to have been used by Ptolemy in formulating this statement. What is significant is that Ptolemy not only does not appear to have copied this statement from Aristotle but, by placing it in a context in which he describes the value of mathematics, he clearly infused the Aristotelian conception of motion with a special mathematical accessibility. Because motion is the key to physics and motion can be described in geometrical terms (in a straight line, in a circle, toward or away from the center), geometry plays a significant role in physics; that is what Ptolemy argued.

Ptolemy's position with regard to the value of mathematics for physics would hardly seem to be based on any statement of Aristotle. Even though Aristotle did not explicitly discuss the usefulness of mathematics for physics, if Ptolemy read through Aristotle's writings he may have formed the opinion that Aristotle would have endorsed the point of view to which Ptolemy himself adhered, namely that mathematics can be valuable in the study of physics. As Thomas Heath has shown, even though Aristotle apparently did not pay any attention to higher mathematics, throughout his works "most of his illustrations of scientific method are taken from mathematics."[12]

Ptolemy stated that not only is mathematics

significant for the study of physics, it would also best prepare the road to the study of the theological nature, because it is the only branch of theoretical philosophy

> able to make a good guess about that activity (ἐνέργεια) which is changeless and separated, because of the kinship of the [mathematical] attributes in those things which are on the one hand perceptible, both moving and being moved, and, on the other hand, eternal and not liable to change, with regard to both the movements and the arrangements of the motions (κινήσεων).

Ptolemy had already located that activity which is the subject of theology in the upper regions of the cosmos. The mathematical attributes of the perceptible, but nevertheless eternal, moving things (presumably, but unnamed by Ptolemy, the heavenly bodies) are akin to that divine activity and can be studied by mathematicians. Therefore, mathematics is the surest path to knowledge of that which is divine and eternal; however, it cannot give true knowledge of the divine, since it does not study the divine activity, but only those kindred attributes which are perceptible.

This is a strong statement; the mathematical study of the celestial bodies is the only form of theoretical philosophy which can give any kind of knowledge, however uncertain, about the subject matter of theology. Ptolemy regarded mathematical astronomy as the best kind of theology which is available to man. He did recognize limitations to man's comprehension of the divine, which he voiced in chapter two of Book Thirteen of the *Syntaxis,* where he cautioned against the inappropriateness of comparing man-made mathematical constructions with the divine. Nevertheless, Ptolemy's claims about the importance of mathematics for theology contrast sharply with Aristotle's views. In the *Metaphysics,* Aristotle explained that mathematics and physics are two branches of wisdom, both of which must depend on the First Philosophy, which he identified as theology.[13] In Ptolemy's view, theology must depend upon mathematics.

As his discussion concerning the dependency of both physics and theology on mathematics shows, Ptolemy was not only concerned with the goal of philosophy, he was also concerned with its methods. Ptolemy stated that "only mathematics can provide sure and unshakeable knowledge to its devotees, provided one approaches it rigorously." He explained that mathematics can provide sure and unshakeable knowledge because its proof proceeds by indisputable methods, that is, arithmetic and geometry.[14]

Ptolemy explained that he was drawn to mathematics, especially to the understanding of the divine and heavenly things, because it is the only branch of theoretical philosophy which is devoted to the investigation of things which are always just as they are. Because mathematics studies such things, it is possible for both its subject matter and for mathematics itself to be always just as they are.

Furthermore, for Ptolemy the value of mathematics was not restricted to the realm of theoretical philosophy. Rather, with regard to "virtuous conduct in practical activities and character," mathematics, "above all things, could make men see clearly; from the sameness, order, symmetry, and calm contemplated about the divinities, making its followers lovers of this divine beauty, accustoming them to a similar state of soul, even as it were forming their natures." Ptolemy came full circle; he began his discussion of philosophy by distinguishing between theoretical and practical knowledge. However, mathematics, for Ptolemy, was not only the best kind of theoretical philosophy, it was also the superior kind of practical philosophy. Man desires to make his soul as similar as possible to the divine; mathematics, while not providing certain knowledge of the divine, nevertheless provides the closest approach to the divine for man. As such, mathematics becomes ethically important.

Ptolemy explained that he thought it fitting to order his own activities with a view to a noble and orderly condition, devoting himself to the teaching of the many beautiful theories ($\kappa\alpha\lambda\hat{\omega}\nu$), especially to those which are particularly called "mathematical." In order to become virtuous and beautiful ($\kappa\alpha\lambda\acute{o}\varsigma$), Ptolemy intended to teach theoretical philosophy, particularly mathematics. This statement is loaded with Platonic overtones.

In his decision to devote himself particularly to theories which are "beautiful," Ptolemy had determined to strive for what Plato considered to be the highest type of human endeavor, that is, the contemplation of the Good and the Beautiful, which are the same things. In the *Symposium,* Socrates recounted at some length the lessons taught to him about the philosophy of Love by a woman named Diotima.[15]

Diotima had explained to Socrates that to love is to give birth to beauty, either physically or spiritually. All people long to propagate in order to gain immortality; those who produce something beautiful of the spirit, for example wisdom, will be able to discuss virtue with others and to teach how it can be achieved. Ptolemy's stated intention to devote himself to the teaching of beautiful theories, with a view to achieving a noble and orderly condition, recalls the views expressed by Diotima. A similar idea is expressed in the *Republic,* in the description of the ascent from the Cave by the would-be philosopher-rulers. Those who have succeeded in the training so far will be brought to the final goal, the ordering of the state and the education of others. Not only must the philosopher discern Beauty for himself, he must teach it to others as well.[16]

It is well to remember here that the *Syntaxis* was conceived not only as a comprehensive treatise, but also as a didactic work. Ptolemy explained in the preface that the work was directed to those who had already advanced somewhat in their studies; even so, he was at pains throughout to explain his methods in detail, so that they could be followed by others.[17] By writing the *Syntaxis,* Ptolemy achieved the goal of the philosopher as stated by Plato; he taught beautiful theories. Furthermore, the very title of the work, the *Mathematical Syntaxis,* may be read with several possible levels of meaning; *syntaxis* can mean treatise or, more literally, "ordering together." The work is a treatise in which he presented his material in an ordered fashion. And in his writing of this work, Ptolemy achieved the Platonic goal for philosophers imposed in the *Republic;* he both created order and educated others.

Once the philosopher has achieved this goal of imposing order and educating others, Plato stated that he would be like a god. In the *Republic,* after the philosopher-kings have fulfilled their obligations of ordering and teaching, "the City shall make memorials and sacrifices for them [the philosopher-rulers] as to divinities if the Pythian oracle approves and, if not, then as to blessed and godlike men." In the *Symposium,* Diotima had explained that when a man has nurtured virtue, he will become the friend of god, and if any other type of immortality is possible for man, he shall have it.[18]

Both these passages from Plato's dialogues speak of an upward climb. In the *Republic,* it is the ascent from the cave; in the *Symposium* it is the climb up the ladder of love. The goal of both is the study of philosophy and the education of others, culminating in the achievement of whatever immortality man is capable of achieving; contemplation of the Beautiful is the way in which man can be most like the gods. This same ideal was spoken of by Socrates in the *Theaetetus* (176b), where he stated that man should become as much like the divine as possible, and that this is accomplished by becoming righteous through wisdom.[19]

This goal of becoming god-like through philosophy was adopted by other philosophers, including Aristotle. In the *Nicomachean Ethics,* he stated that by following the life of theoretical philosophy, man will be happiest and will be blessed. He explained that man is most happy when he is like the gods and offered advice for how to achieve this goal:

> It is not necessary, as according to some advisors, for human beings to think about human things, or for mortals to think about mortal things, but rather as much as possible to be immortal and to do all things, with regard to living, according to the best in oneself.

But Aristotle's formulation of how to become god-like, namely living the contemplative life, was not that followed by Ptolemy. For Plato and for Ptolemy, teaching was necessary for the achievement of the ideal. This emphasis on education was also explicitly stated by the author of the Platonic commentary known as the *Didascalicos,* who explained that we may become like the gods in several ways, but most legitimately through reason and teaching and the transmission of theories.[20]

Ptolemy had, so far, not painted a self-portrait of an Aristotelian. Indeed, throughout his discussion of the division of philosophy his choice of words and phrases hints at various philosophical influences and sources for his ideas. Certainly, the influence of Aristotle is present in Ptolemy's writing but, clearly, he did not subscribe to Aristotle's division of philosophy. The influence of Plato and of Platonic doctrines as expressed by the Middle Platonists is present in Ptolemy's preface.

Given Ptolemy's clear deviation from Aristotle's views, it is puzzling that the only philosopher he mentioned by name was Aristotle. Even though he disagreed with Aristotle, Ptolemy never said that Aristotle was misguided in his division of philosophy. It seems likely that contemporary readers of the *Syntaxis,* who, one presumes, would have been interested in mathematics, would have been sympathetic to the Platonic overtones of the preface. Ptolemy didn't need to cite the authority of Plato by name; perhaps his mention of Aristotle is an indication of a conciliatory and syncretic approach to philosophy. What about the evidence of "Stoic" influences in his thought?

As part of the distinction he drew between theoretical and practical philosophy in the beginning of the preface, Ptolemy contrasted the way in which the most benefit is achieved in each; in practical philosophy it is achieved through continual activity in practical things, while in theoretical philosophy the most benefit is gained "from the progress of theories." This idea of progress contains Stoic overtones, for προκοπή (progress) was one of their catch-words. However, in Stoic philosophy, discussions of progress (προκοπή) are always in the context of ethics. The term προκοπή was not reserved for use by Stoics; the *Didaskalikos* described most men as being in a state of progressing (προκόπτοντας εἶναι) between moral excellence and meanness.[21] What might at first glance look like an indication of Stoic influence need not be interpreted as such.

Ptolemy's use of the word προκοπή appears at first reading not to have been made in an ethical context. However, when he turned to the discussion of his own motivation for studying theoretical philosophy, this ethical undertone becomes clearer.

Ptolemy stated that he would attempt to accomplish his goal in the *Syntaxis* by pursuing those studies

> which have already been mastered by those who approached them in a genuine spirit of enquiry and by ourselves attempting to contribute as much advancement as has been made possible by the additional time between those people and ourselves.

Ptolemy acknowledged that his work would be to some extent based on the work of his predecessors. But he was particularly concerned with making his own contribution, to mark his own progress. He was not content with summarizing the achievements of his predecessors, but had to break new ground. For this is precisely how one achieves the most benefit from theoretical philosophy, as Ptolemy had already asserted.

Returning to the distinction between theoretical and practical philosophy made at the beginning of the preface, it has already been noted that, for Ptolemy, mathematics qualified both as practical and theoretical philosophy. For Ptolemy, studying and teaching mathematical theories was an ethical endeavor. Ethical virtues may be had without education, but the greatest ethical virtue, knowledge of the theory of the universe, is obtained only through education, which requires teaching. Since mathematics is the best philosophy, by teaching the mathematical theory of the universe, Ptolemy demonstrated that he had not only obtained that greatest of virtues, he also enabled others to achieve it as well. The goal of man is to become as like the divine as is humanly possible, which can be done through mathematics. For Ptolemy, the best philosophy, that is, mathematics, is a quasi-religious way of life.

Throughout his preface, Ptolemy's ambitions and goals are those of the philosopher, albeit Ptolemy's conception of what constitutes a philosopher is somewhat idiosyncratic. Ptolemy's conception of philosophy is neither Platonic nor Aristotelian; much of what he says about philosophy is a denial of the possibility of achieving the philosophical goals of Plato and Aristotle. Nevertheless, in his own view, Ptolemy was a philosopher.

Certainly the *Syntaxis* as a whole is devoted to mathematics. But Ptolemy made his position clear within his short philosophical preface: he regarded mathematics as the First Philosophy. By studying and teaching mathematics, Ptolemy considered himself to be working toward the highest goal of the ancient philosophers—the achievement of the sort of immortality appropriate to man. To become like the eternal divinities through mathematics, which is concerned with the eternal celestial divinities, was Ptolemy's desire and goal. This desire to be as similar to the immortal gods as is humanly possible was not only voiced by Plato and Aristotle but, in fact, actually seems to have become an even more urgent goal of the Hellenistic philosophers. The quasi-religious character of Ptolemy's conception of philosophy, as stated in Book One of the *Syntaxis,* has much in common with that of other Hellenistic philosophers, including the Stoics and Epicureans.

Ptolemy's desire to share in the divine was in no way unusual. That a philosopher could gain a reputation as a god in the Hellenistic period was demonstrated by the description of Epicurus of Samos given by Lucretius: "he was a god, indeed a god, . . . who first invented that principle of life now called philosophy."[22] Ptolemy's First Philosophy was mathematics; in this respect he was not Aristotelian. In his desire to achieve immortality through philosophy he was thoroughly Hellenistic. And, of course, in one sense at least, he achieved his goal.

✳ 3
The Hypotheses Underlying the *Syntaxis*

Following the outline of his views on the organization and philosophy of science presented in the first chapter of the *Syntaxis,* Ptolemy summarized the contents of his book in the second chapter. He stated that he would begin the treatise with a view of the general condition of the Earth as a whole with regard to the heaven as a whole. He explained that, in general, these things should be assumed in advance:

> both that the heaven is spherical in shape and moves spherically and that the Earth itself is also, to the senses, spherical in shape, when taken as a whole, and in position lies in the middle of the entire heaven very much like the center, and in size and in distance has the ratio of a point to the sphere of the fixed stars, [the Earth] itself having no motion involving change of place.

Ptolemy clearly expected his readers to be familiar with these ideas, noting that he would only discuss them briefly, to serve as a reminder. He did this in chapters three through seven of Book One. Ptolemy referred to these five points as "hypotheses" ($ὑποθέσεις$) and made it clear that he considered discussion of these points to be necessary as an introduction to the discussion of particular topics and what follows.[1]

It is only at the beginning of chapter eight that Ptolemy introduced another idea which is also to be presupposed, namely, that there are two primary motions in the heavens. Ptolemy did not refer to this presupposition explicitly as an hypothesis; he merely stated that as a general assumption, it should be added to the others already mentioned. It is possible that Ptolemy was subtly signalling a somewhat different explanatory status for this final presupposition. However, if that was his intention, the functional difference between this presupposition and those mentioned earlier is not made clear by Ptolemy. Furthermore, at the beginning of chapter nine, he indicated that these presuppositions all share the same status. Accordingly, these six points which are necessary to be established or assumed before proceeding on to the other matters will here be called hypotheses. Ptolemy stated that it was sufficient to treat the hypotheses briefly at the outset, because they would be confirmed by the agreement between the subsequent demonstrations and the phenomena.[2]

Toomer discussed some of the problems involved in understanding Ptolemy's use of the word $ὑπόθεσις$ (hypothesis). He explained that

> with some hesitation, I have used this [hypothesis] to translate ὑπόθεσις, although the connotation in the Almagest never really coincides with the modern one. Whereas we use 'hypothesis' to denote a tentative theory which has still to be verified, Ptolemy usually means by ὑπόθεσις something more like 'model', 'system of explanation', often indeed referring to 'the hypotheses which we have demonstrated'. The word still retains much of the etymological meaning of 'basis on which something else is constructed'.³

Thus, the word "hypothesis" has, etymologically, a meaning of "foundation" or "something fundamental."

Both Plato and Aristotle discussed the status of hypotheses as the foundation of mathematical explanations. Plato, in the account of the Divided Line given in the *Republic,* described the hypotheses used by mathematicians as assumptions for which no account is given. In dialectic, such assumptions would be treated not as beginning principles, but as a springboard to the *archai,* which are the true starting-points and require no assumptions. Here Plato was discussing dialectic, but it is clear from his example that mathematicians of his time treated hypotheses as assumptions which needed no justification. For Plato, these hypotheses must depend ultimately on the *archai.* The way in which mathematicians proceed from assumptions is not the proper way for philosophers to proceed, as Socrates argued in the *Meno.* Jacob Klein pointed out that in the *Republic* Plato was "playing with the meaning of the term ὑπόθεσις." He noted that whatever the range of the "technical" meanings of the word (in, for example, mathematical and medical treatises), "they all imply something *without* which something else cannot be or cannot be conceived." The problem is that, in contrast with "that which depends, for its being or its being conceived, on a foundation or 'supposition' (not necessarily a conjectural one), the foundation takes *precedence.*" Plato questioned the nature of the precedence of hypotheses, discarding such assumptions as the basis of philosophical inquiry.[4]

Turning from mathematics to logic, in Book One of the *Posterior Analytics,* Aristotle explained the difference between a definition (ὁρισμός) and an hypothesis (ὑπόθεσις): an hypothesis assumes that something either is or is not, while a definition states what something is. Aristotle did not suggest that the statement of an hypothesis, as such, required any sort of justification. With regard to mathematical hypotheses, Aristotle clearly believed that First Philosophy itself must account for the principles of mathematics.[5] For Aristotle, the account of these principles would necessarily come from philosophers, rather than mathematicians.

Although none of the common principles of the extant text of Euclid's *Elements* are called hypotheses, this seems to have been one name for what we call the ὅροι, or definitions. In his fifth-century commentary on the first book of the *Elements,* Proclus divided the common principles (κοιναὶ ἀρχαί) as enunciated by Euclid into hypotheses, postulates, and axioms (ὑποθέσεις, αἰτήματα, and ἀξιώματα), instead of using the terminology in our extant *Elements* of definitions, postulates, and common notions (ὅροι, αἰτήματα, and κοιναὶ ἔννοιαι).⁶ What we know as a definition in our version of the *Elements* was apparently labeled an hypothesis in the *Elements* read by Proclus.

Proclus discussed the terminology, and made it clear that because "no science demonstrates its own first principles," definitions or hypotheses need no justification. The extended discussion which Proclus provided on the use of such terms as "hypothesis" (he noted that axiom, postulate, and hypothesis are sometimes all called hypotheses) indicates that there was not a single universally accepted usage. (He mentioned, presumably as an example of the variety of usage, that the Stoics called every simple statement an axiom.)⁷ It is not, however, clear from Proclus when this confusion of terminology began.

It seems clear from Plato, Aristotle, and Proclus that hypotheses were not customarily justified by mathematicians. In the first nine chapters of the *Syntaxis*, Ptolemy's use of the word "hypothesis" and his discussion of the hypotheses as prerequisite to the remainder of the work indicates that he envisioned the physical hypotheses as starting-points. But, in providing a discussion of each hypothesis, he implies that they can, and must be, justified in some way; simply to assume and state the hypotheses is not sufficient.⁸ Certainly, he did not share the dissatisfaction of Plato with the use of hypotheses, nor did he follow Aristotle and suggest that the hypotheses must be explained by metaphysics. Nevertheless, Ptolemy was not content to function as the typical mathematician described by Plato and Aristotle, adopting assumptions as hypotheses without further consideration. Ptolemy's attitude toward the adoption of hypotheses is that even if they cannot be demonstrated, their adoption must at least be justified.

For this reason, here the term "hypothesis" must be understood to have a meaning somewhat idiosyncratic to Ptolemy. The word ὑπόθεσις is used throughout the *Syntaxis* to refer to the hypotheses for uniform circular motion, the epicyclic and eccentric hypotheses, first introduced in chapter three of Book Three. Toomer suggested that in the *Syntaxis* "hypothesis" can mean "model" or "system of explanation," as in 13.1, where the eccentric hypothesis and the epicyclic hypothesis are discussed. While it should be noted that Ptolemy does sometimes use the term "hypothesis" differently in different contexts, there is always the sense that the hypotheses (be they "assumptions" or "models") can be shown to be in agreement with the phenomena.[9]

Furthermore, it will become evident from the following examination of the physical hypotheses that the justification of an hypothesis was, for Ptolemy, based on agreement with empirical observation.[10] That an hypothesis is liable to and even requires a justification based on empirical experience seems to have been peculiar to Ptolemy; the *archai* upon which Plato would have hypotheses depend would surely not have been empirical, nor would the First Philosophy of Aristotle have demanded an empirical account of hypotheses.

The importance of accounting for the phenomena was underscored again in Book Nine of the *Syntaxis*, where Ptolemy explained that those things which are assumed without proof (ὑποτιθέμενα), when they have been conceived in such a fashion as to be consistent with appearances, rely on some plan and knowledge, even though the manner in which they were arrived at is difficult to ascertain.[11] For Ptolemy, consistency with the phenomena was a basic requirement for the adoption of an hypothesis. While the basis of arriving at a particular hypothesis may have been impossible for Ptolemy to explain, its agreement with the phenomena can be demonstrated. Ptolemy provided such demonstrations in the following six chapters of the *Syntaxis*.

I. *That the heaven moves spherically*[12]

Ptolemy wrote that the ancients may have been led to the idea that the heaven moves spherically by observations of the celestial bodies rising and setting in parallel circles, motion that recurs. He believed that this idea, once obtained from such considerations, accorded with all the appearances. Further, the appearances contradict all alternative ideas. In support of this viewpoint, Ptolemy offered arguments based upon many types of considerations: astronomical, optical, chronological, mathematical, and physical.

He first attacked the idea that the celestial bodies move in straight lines toward infinity (ἐπ' ἄπειρον) and pointed to problems that cannot be resolved by this idea. First, the stars are observed to move from the same starting point every day.

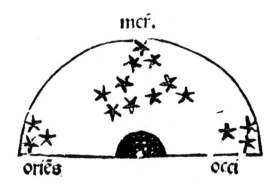

Figure 3.1 The rising and setting of stars ([Sacro Bosco], [*Sphaera mundi*], [Venice], Erhard Ratdolt, 1485, [p. 4]).

Ptolemy asked, if the stars are travelling toward infinity, how could they turn back, especially since they do not appear to do so? Second, if the stars are moving toward infinity, their size would gradually diminish toward the vanishing point as they get farther away; yet this does not accord with the phenomena, for they seem to be larger at the moment of their disappearance, at which time they are cut off by the surface of the Earth.[13]

Note how Ptolemy's objections were based on his knowledge of the appearances; he would not accept an hypothesis that required ignoring the phenomena. Ptolemy's arguments against this hypothesis were not based on a physical theory.

Although Aristotle had disallowed motion toward infinity in his own physical system,[14] Ptolemy was not following him in his rejection of this explanation. Ptolemy objected to the idea that the celestial bodies move in infinite straight lines on the basis of empirical considerations; Aristotle had rejected the same idea for philosophical reasons. Furthermore, Aristotle's arguments

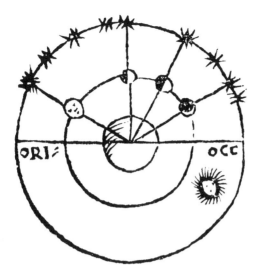

Figure 3.2 The circular motion of the stars, Sun, and Moon (Sacro Bosco, *Sphaera,* Lyons, 1564, p. 10).

against motion toward infinity were presented as part of a larger argument, that there cannot be a plurality of worlds.

That all change occurs within fixed limits was the basis of Aristotle's objection to the idea that motion could occur toward infinity. He illustrated this principle with the examples that a patient being cured is between sickness and health, and that anything which is growing is between smallness and greatness. Such limits apply to local motion as well, for which there must be an end (τέλος); motion cannot be without limits (μὴ εἰς ἄπειρον φέρεσθαι).[15]

To corroborate this assertion Aristotle argued that the element earth moves more quickly the closer it is to the center and fire more quickly the farther upward it is. If motion were to infinity, speed would also be infinite; if speed were infinite, so would be weight or lightness.[16] Earlier in the same work he had offered a proof that for a body to have infinite weight (or infinite lightness) is impossible, based on considerations of the amount of time it would take an infinitely heavy or light body to move.[17] This argument is based on the internal consistency of Aristotle's own natural philosophy, not upon an appeal to observation. It is part of his larger argument against the existence of infinite bodies.

In the sixth century, Simplicius commented on a related passage from *On the Heavens* (277a27) and admitted some skepticism concerning the doctrine of natural place, particularly with regard to Aristotle's discussion of weight. He stated that Ptolemy had performed experiments to contradict Aristotle's claims and reported that these experiments had been replicated by others,

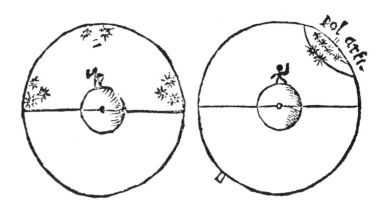

Figure 3.3 Stars that rise and set (left). Circumpolar stars (right) (Sacro Bosco, *Sphaera,* Lyons, 1564, p. 12).

including himself, with similar results, contradicting the claims of Aristotle.[18]

Ptolemy also objected to the idea that stars rise up from the Earth and then fall back toward it. Ptolemy found this suggestion to involve at least one of any number of absurd notions: either that random chance could account for the order ($τάξις$) of the stars; or that part of the Earth has an incandescent nature that ignites the stars, while another part extinguishes stars; or that the same part of the Earth can both ignite and extinguish the stars. He appealed to observation in his refutation of this suggestion, stating that it cannot account for the always-visible (circumpolar) stars, or the phenomenon that different stars are visible from different places.

Ptolemy suggested that if the motion of the heavens were not spherical, the distance of a star from the Earth would vary during the course of each revolution. He discounted the larger appearance of stars when they are near the horizon as evidence of closer proximity, claiming that this appearance is due to an optical effect, that is, the steaming up of the moisture surrounding the Earth between them and our eyes. He compared this effect to that of objects which appear larger the farther down they are immersed in water.[19]

Additionally, Ptolemy offered several terse reasons, for which he offered no support, for declaring the motions of the heaven to be spherical. First, he claimed that only the hypothesis of spherical motion accounts for the successful construction and use of *horoskopia,* which here appears to be a general term for instruments used for the measurement of time.[20]

Another reason to favor the hypothesis that the heaven moves spherically was based on the

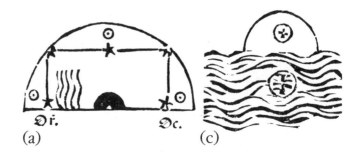

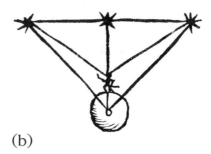

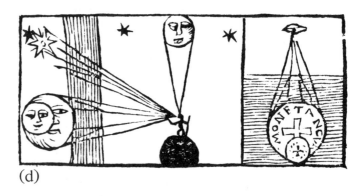

Figure 3.4 (a, b) Stars do not move in straight lines. (c, d) That stars appear larger near the horizon is due to an optical effect ([*Sphaera mundi*], [Venice], Erhard Ratdolt, 1485, [p. 6] [*top*]; Sacro Bosco, *Sphaera,* Lyons, 1564, p. 17 [*bottom*]).

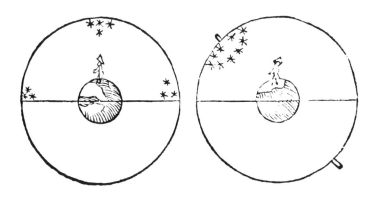

Figure 3.5 Stars that rise and set (left). Circumpolar stars (right) (Francesco Barozzi, *Cosmographia in qvatvor libros,* Venice, 1585, p. 13).

assumption that "the movement of the heavenly bodies is the most unhindered and most well-moving" (ἀκωλύτου τε καὶ εὐκινητοτάτης). Ptolemy stated that the circle among plane figures offers the easiest path of motion, as does the sphere among solids.

Aristotle, in the *Physics,* had offered four arguments for the primacy of circular motion. Following Cornford, the four arguments may be summarized as follows: (1) it is simpler, more complete, and capable of being eternal (265a13–27); (2) there is no definite starting- or end-point to the motion of a sphere; it does not change its place as it rotates, but moves continuously (265a27–b8); (3) it is the measure of all motion and change (b8–11); (4) it is the only motion capable of being uniform (b11–16). Ptolemy did not employ any of Aristotle's arguments, perhaps feeling they would be known to his readers. In any case, Ptolemy did not argue for his position, an indication that he thought that his judgment would be unquestionably shared by others.

Ptolemy also offered a geometrical reason for adopting the hypothesis of the spherical motion of the heavens, based perhaps on the work of Zenodorus on isoperimetric figures, perhaps completed in the early second century B.C. Ptolemy stated that "since of different figures having equal perimeters those having more angles are greater [in area], the circle is greater than plane figures and the sphere greater than solids." Ptolemy's geometrical justification of the spherical motion of the heavens was based on the sphere having the greatest volume compared to figures of equal perimeter. Zenodorus had proven this proposition in *On Figures of Equal Boundary* (Περὶ ἰσοπεριμέτρων σχημάτων) and Ptolemy may have been familiar with his proof.[21] Ptolemy concluded this line of thought by stating that "the heavens are greater than other bodies." Although Ptolemy was not clear at this point, he seems to have meant that since the heavens encompass all other bodies and have the greatest volume, the heavens must therefore be spherical.

Ptolemy then offered a slightly more developed argument, although he was still laconic here. What follows is based on what he referred to as "certain physical things" (ἀπὸ φυσικῶν τινων), but in fact was founded on both mathematical and physical considerations. First, he stated that of all bodies, the aether (αἰθήρ) has finer parts (λεπτομερέστερος) and is more homoeomerous (ὁμοιομερέστερος), that is, has more similar parts.

His use of the word λεπτομερέστερος (having smaller, finer, more rarified parts) to describe the aether indicates that Ptolemy was thinking of the aether in physical terms. The word λεπτομερέστερος was used by Aristotle in his discussions of the theories of those earlier philosophers who stated that fire was the primary element. Of course, in these passages, Aristotle was criticizing the theories of his predecessors. Furthermore, neither passage is concerned with the aether. Book Three of *On the Heavens* is concerned not with the primary body (τὸ πρῶτον τῶν σωμάτων), the aether, but rather with those elements which are subject to generation.[22] The passage from *On the Soul* is a consideration of earlier ideas regarding the nature of the soul.

What is particularly striking about these passages, one found in *On the Heavens,* the other in *On the Soul,* is that they deal with the same characteristics that Ptolemy used to describe the aether. In the first, the notion of homoeomerity occurs and in both the question of the shape of the constituent parts of the elements arises; both the homoeomerity and the shape of the parts of the aether are discussed by Ptolemy in the passage under consideration. Furthermore, in the passage from *On the Soul,* the idea of the divinity of the celestial bodies is mentioned, an idea to which Ptolemy alluded in his own discussion.

The two words used by Ptolemy to describe the aether seem to emphasize its particular nature; both λεπτομερέστερος and ὁμοιομερέστερος refer to the parts of the aether, which are on the one hand smaller, and on the other hand more similar, than all other bodies. Several modern readers have gone so far as to ascribe to Ptolemy a "molecular" conception of matter, without considering the anachronistic problems involved in this term.

The word ὁμοιομερής (homoeomerous) has significance in physical theory, being particularly associated with the ideas of Anaxagoras of Clazomenae (fifth century B.C.), as well as with Aristotle's own theory of matter.[23] Book Four of the *Meteorology,* probably a Peripatetic work if not written by Aristotle himself, discusses the homoeomerity of substances at some length. These associations might lead us to assume that it must have physical significance here. However, what Ptolemy said about homoeomerous bodies, that they have surfaces which are homoeomerous, seems to be peculiar until we realize that the word ὁμοιομερής also had a specific mathematical usage.

Ptolemy's use of the word in this passage may be the earliest extant example of its employment with a mathematical meaning. But there are indications that it may have been used in a special geometrical sense before Ptolemy, and that would help to explain why he does not elaborate further than his statement that homoeomerous bodies have surfaces with homoeomerous parts. He may have assumed that his readers, presumably mathematicians, would be familiar with the mathematical usage of the word.

Zenodorus, as reported by Theon of Alexandria (fourth century A.D.) in his commentary on the *Syntaxis,* described certain geometrical figures as homoeomerous: "both the circle, of plane figures, and the sphere, of solid figures, are homoeomerous." Zenodorus's work is no longer extant; although fragments are preserved in three sources, not much is known about him, and it seems unlikely that a full understanding of his conception of geometrical homoeomerity can be gained. Ptolemy may have been familiar with Zenodorus's description, for his next statement is that "of plane figures, only the circle, while of solid figures, only the sphere, display homoeomerous parts," which is very similar to the Zenodorian fragment.[24]

It is only in a late work, namely Proclus's *Commentary on Euclid,* that a definition of homoeomerous lines and surfaces is extant. In his discussion of the definition of the line given by Euclid, Proclus, succinctly defined the homoeomerous line as "a line having all of the parts fitting each other," either a straight line, a circular line, or a cylindrical spiral. However, Proclus made it clear that these were not recently defined, for he cited Apollonius as having proven the homoeomerous property of the cylindrical helix (the *kochlias* or *kochlion*) in a now-lost work. Furthermore, Proclus credited Geminus with having proven that "if two straight lines drawn from a point to a homoeomerous line make equal angles to it, the straight lines are equal," and that there are only three homoeomerous lines.[25]

In his explanation of Geminus's classification of lines, Proclus stated that "there are only three lines that are homoeomerous: the straight line, the circle, and the cylindrical spiral." Additionally, he noted that of the lines that are on the surface of solids, "some are homoeomerous, like the spiral around a cylinder, but the others are anhomoeomerous." He also explained that there are only two homoeomerous surfaces, one being plane, the other spherical. The cylinder is not a homoeomerous face because all of the parts of the cylinder do not fit all of the other parts.[26] Thus, the sphere is the only homoeomerous three-dimensional surface.

Ptolemy assumed that the aether was homoeomerous, an assumption which may have been based on a physical theory, since he stated that this argument was based on physical considerations. But even though he made this assertion, he then went on to make statements about geometrical figures. It is on the basis of the geometrical property of being homoeomerous that he concluded that since the aether is a solid figure, it must be spherical, because a sphere is the only three-dimensional figure which is homoeomerous. But the principle of being geometrically homoeomerous, according to Proclus, applies only to the surfaces of homoeomerous figures, not to chunks of physically spherical bodies. Nevertheless, Ptolemy applied this conception of homoeomerity not only to geometrical figures, but to matter itself.

Furthermore, Ptolemy used the geometrical conception of homoeomerity to distinguish between celestial and terrestrial matter. He contrasted earthly, corruptible bodies ($\sigma\omega\mu\alpha\tau\alpha$) to the divinities in the aether ($\tau\grave{\alpha}\ \dot{\epsilon}\nu\ \tau\hat{\omega}\iota\ \alpha\dot{\iota}\theta\acute{\epsilon}\rho\iota\ \kappa\alpha\grave{\iota}\ \theta\epsilon\hat{\iota}\alpha$), by which he means the celestial bodies. Furthermore, he did not state that the divine bodies are composed of aether, but described them as *in* the aether. The earthly bodies are constituted wholly from rounded ($\pi\epsilon\rho\iota\phi\epsilon\rho\hat{\omega}\nu$), anhomoeomerous shapes ($\sigma\chi\eta\mu\acute{\alpha}\tau\omega\nu$), while the divine things in the aether are from homoeomeries and spheres ($\dot{o}\mu o\iota o\mu\epsilon\rho\hat{\omega}\nu\ \kappa\alpha\grave{\iota}\ \sigma\phi\alpha\iota\rho\iota\kappa\hat{\omega}\nu$). There is an imprecision in the shape of earthly bodies, while the heavenly bodies are both geometrically precise and are mathematically unique, being spheres, the only homoeomerous solid figure. This, then, is a purely *mathematical* treatment.

However, Ptolemy stated at the outset that his view was also supported by *physical* considerations. He did state that the aether is not only homoeomerous, but, of all bodies (σωμάτων πάντων) it is *more* homoeomerous. Thus he must be speaking physically, since mathematically a line or a surface is homoeomerous or it is not. Further, the use of the word σώματα for both earthly and corruptible things and for the aether indicates that he considered the aether to be both geometrically and physically homoeomerous.

The word homoeomeries (τὰ ὁμοιομερῆ) was used by Aristotle in connection with his discussion of Anaxagoras's theory of matter. Aristotle himself used homoeomeries to describe those natural substances for which every part is exactly like the whole, such as skin, hair, and bone. Indeed, Aristotle used the term homoeomeries most often in reference to the substances of which animals are composed. In the *Physics* he used the term to refer to Anaxagoras's matter theory in his discussion of whether there is an infinite number of elements. Once he had shown to his own satisfaction that there is not, he then used the word in his own classification of matter, referring to the natural substances listed above, as well as in a discussion about the material elements.[27]

As mentioned above, Ptolemy's use of the term "homoeomerous" is the earliest extant use in the mathematical context, although there are earlier examples of "homoeomerous" having been used in the physical sense. Zenodorus, or whoever it was that first applied it to geometry, may have simply borrowed a word which meant "having uniform parts" and given it a specific mathematical meaning.[28] Ptolemy, however, seems to have combined the separate physical and mathematical meanings of homoeomerous, as is demonstrated by his dependence on the geometrical usage of the word in a passage which he himself labeled as being based on physical considerations.

In fact, Theon of Alexandria, in his *Commentary*, emphasized that Ptolemy was writing in physical terms. He paraphrased Ptolemy's words, stating that the aether is a physical, homoeomerous body, and is at the same time a physical (or natural, $\phi\upsilon\sigma\iota\kappa\acute{o}\nu$), homoeomerous body which is spherical, possessing a homoeomerous surface. Therefore the aether is spherical.[29]

What we see here, the combination of a physical notion with a mathematical definition, may have originated with Ptolemy, or, perhaps, an earlier writer. Ptolemy's off-hand use of the word "homoeomerous" in a mathematical sense would suggest that the adoption of "homoeomerous" as a mathematical term was pre-Ptolemaic; it may simply have been done on the basis of analogy.

Furthermore, because of his views regarding the less-than-certain character of physics, it would seem surprising if Ptolemy had adopted a physical concept such as homoeomerity and applied it to mathematics. But he did use the term in *both* the mathematical and physical senses, indicating that he was aware of both meanings. Perhaps the mathematical meaning of the term "homoeomerous" seemed to him to be primary, particularly since it was liable to geometrical proof, being the name for a property that is defined, then proved for particular cases.

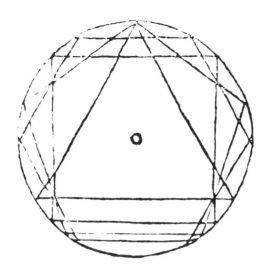

Figure 3.6 The circle is greater than any other plane figure (Sacro Bosco, *Sphaera*, Lyons, 1564, p. 14).

To summarize, Ptolemy's justification for the circular motion of the heaven is quite different from that offered by Aristotle for the same idea. Aristotle assumed the circular motion of the heaven, basing this assumption on considerations deriving from his theory of motion, from which, in turn, he derived his theory of matter. In the *Physics*, he made it clear (185a12–13) that his first assumption about nature was its motion, having stated that the things which exist by nature are, either all or some of them, in motion. He explained (200b12–15) that since nature is a principle of motion and change, the method for studying nature must consider motion and change. And, following his own methodological bent, a great deal of the *Physics* was concerned with the definition and explanation of motion and change. For Aristotle, local motion was the primary kind of motion, of which circular motion is the primary type. Circular motion is the only regular motion, and is single, continuous and infinite. In *On the Heavens,* the elements are distinguished by their motion; Aristotle explained that since the vertical motions belong to the four elements, there must be another element which enjoys circular motion, the aether.[30] Thus, the circular motion of the celestial element, aether, forms an important part of his treatment of motion, which for Aristotle is the cornerstone of his physical theory. Circular motion is equally important in Ptolemy's discussion; however, the mathematical character of his treatment distinguishes it from that of Aristotle.

II. *That the Earth, taken as a whole, is sensibly spherical*

In the fourth chapter of Book One of the *Syntaxis,* Ptolemy provided two arguments for the sphericity of the Earth, both ostensibly based on observations. The extended first argument depends on celestial phenomena; the second, which is very brief, depends on terrestrial phenomena. In both arguments Ptolemy's preoccupation with mathematics is evident; the analogy of the sensibly spherical Earth to a geometrical model of a sphere underlies the discussion of this hypothesis.

Beginning his discussion with a consideration of celestial observation, Ptolemy stated that the Sun, Moon, and other stars are seen not to rise

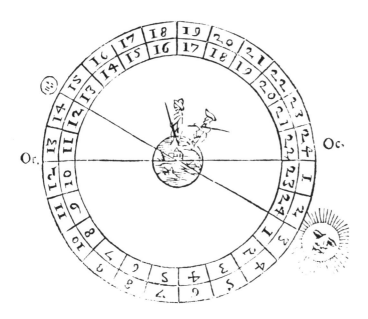

Figure 3.7 Eclipses of the Moon occur simultaneously for all observers, but not at the same hour (Francesco Barozzi, *Cosmographia in qvatvor libros,* Venice, 1585, p. 21).

and set simultaneously for everyone on Earth, but rather always earlier for those more towards the east, later for those towards the west. In contrast to this, he noted those phenomena which occur simultaneously for everyone, that is, eclipses, particularly those of the Moon, do not occur at the same hour (that is at an equal interval from noon) for all recorders; rather the hours recorded by those observers in the east are always later than those recorded by observers in the west. Further, "the differences in hour are analogous [in proportion] to the distances of the places."

He rejected other possible shapes for the

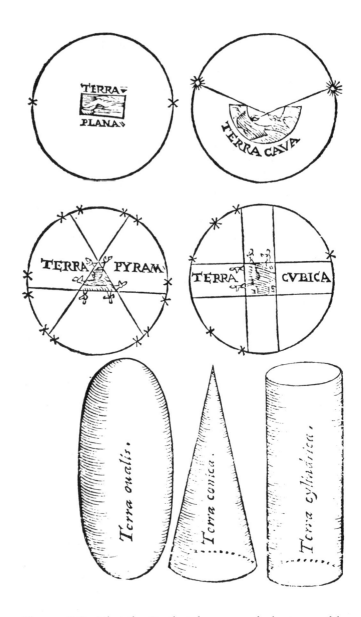

Figure 3.8 That the Earth, taken as a whole, is sensibly spherical. Illustrations of the impossible consequences of other shapes. (Francesco Barozzi, *Cosmographia in qvatvor libros,* Venice, 1585, p. 22).

Earth (concave, plane, polygonal, and cylindrical) for the reason that celestial phenomena would not appear as they do if the Earth were any shape other than spherical. He explained that if the Earth were concave

> the rising stars would appear first to those in the west, but if it were plane, they would rise and set at once and at the same time for everyone on the Earth, and if it were triangular or square or some other polygonal figure, they would again appear in the same way to those living on the same straight side.

Yet, as he noted, this is not what happens.

In particular, he raised objections to the possibility of a cylindrical shape for the Earth, which the observations concerning differences in the time of the rising and setting of stars would seem to allow if the ends of the cylinder were at the poles of the universe, noting that, if that were the case none of the stars would be always visible, rather, all stars would rise and set or would always be invisible for all observers. Against this possibility he noted that when travelling north, southern stars disappear and more northern stars become visible. The idea is that the curvature of the Earth blocks the view for observers on Earth. From these considerations, Ptolemy concluded that the surface of the Earth is spherical.

Turning from celestial observations to terrestrial, Ptolemy briefly mentioned that when sailing toward mountains or elevated places, they appear to rise gradually from the sea; this is due to the curvature of the surface of the water on Earth. That he provided such a cursory statement about terrestrial phenomena indicates that he considered the celestial phenomena to be more important in justifying the hypothesis concerning the sphericity of the Earth.

Ptolemy did not cite any physical justifications for the hypothesis concerning the Earth's sphericity. The evidence he did offer is strikingly different from that which was proffered by Aristotle.

In chapter thirteen of the second book of *On the Heavens,* Aristotle had noted that there had been much disagreement about the shape of the Earth, with some believing it to be spherical, and others suggesting a drum shape. He rejected the observational evidence for the latter view (that the rising and setting Sun displays a straight instead of a curved line where it is cut off from view by the horizon, whereas if the Earth were spherical, the cut would by necessity be curved), and complained that those who adhered to the drum-shape view failed to consider the distance of the Sun from the Earth and the size of the Earth's circumference. He argued that it appears to be straight because it is manifesting its appearance on a small circle very far away, and concluded that the phenomena do not necessitate a disbelief in a spherical Earth.[31]

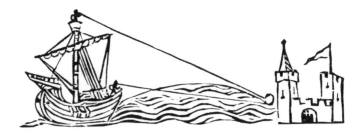

Figure 3.9 The Earth is spherical ([*Sphaera mundi*], [Venice], Erhard Ratdolt, 1485, [p.7]).

The remainder of chapter thirteen is devoted to a discussion of whether the Earth is in motion or at rest, and its central position in the universe. At the end of the chapter Aristotle stated that the discussion of the shape, place, and motion of the Earth was concluded. But, in chapter fourteen, Aristotle returned to the question of the shape of the Earth, and stated that it must be spherical. He offered two arguments based on the idea of natural place and two arguments based on astronomical observations to support this claim.

In the first argument concerning natural place, Aristotle suggested a thought-experiment concerning the generation of the Earth. Assuming that it is the nature of the heavy to move toward the center, he suggested we imagine that the generation of the Earth occurred as some of those who have reasoned about nature described. Aristotle suggested that, when the mixture was merely potential, the various parts moved toward the center. Explaining that any body endowed with weight, no matter what size, moves toward the center, Aristotle stated that rather than stopping when its edge meets the center, a body will continue until its own center occupies the center. He concluded that if the Earth was generated, it must have been formed in this way, and so it is clear that its generation was spherical. But, he argued, even if the Earth is ungenerated, it has the same character just as if it had been, at the beginning, generated.[32]

Aristotle's second argument for the spherical shape of the Earth (*On the Heavens* 297b19–21), based on the doctrine of natural place, is based on his assertion that heavy bodies move toward similar, rather than parallel, angles. As Guthrie pointed out in the notes to his edition and translation, Aristotle's meaning is not completely clear. Aristotle had used the same phrase (πρὸς ὁμοίας γωνίας) earlier in the chapter. Stocks explained the phrase as meaning that the angles at each side of the line of fall of any one body are equal. Guthrie questioned this reading, asking whether it does "not more naturally mean that the angles made by one falling body with the earth are similar to those made by another?"[33]

Aristotle continued by noting that either the Earth "is spherical or is spherical by nature."[34] He noted that it is proper to refer to things according to their nature, rather than to refer to them with reference to that which constrains them. According to this line, because the Earth is spherical by nature, it is proper to call it spherical.

However, two arguments based on astronomical observations were also offered by Aristotle to corroborate his philosophical argument. He contrasted the concave edge of the Moon during eclipses with the other shapes it exhibits during its monthly phases: straight-edged, gibbous, and concave. He argued that if eclipses are due to the position of the Earth, the Earth's shape must be spherical, otherwise the Moon would not exhibit only a concave edge during eclipses, as it does.[35]

That Ptolemy did not use the argument of the shape of the shadow cast on the moon as evidence of the Earth's sphericity has been noted with some wonderment by Pedersen. However, as Neugebauer has pointed out, this argument is mathematically inconclusive for a number of reasons, including the problem of accurately establishing the nature of the observed curve.[36]

On the other hand, Ptolemy's own discussion

is somewhat reminiscent of Aristotle's second astronomical argument, which is based on the observation of stars. Aristotle noted that different stars are seen in the north than in the south. Certain stars are seen in Egypt and around Cyprus, but are not visible in places towards the arctic, and the stars which are continually visible in the north are stars which set in other places.[37] Aristotle presented these observations in a rather casual and commonplace way, suggesting that they were knowledge shared by many. Ptolemy's discussion was similarly off-hand, perhaps also relying on observations familiar to many readers.

The assumption of a spherical Earth is fundamental to Book Two of the *Syntaxis,* which is devoted to a detailed discussion of the κλίματα (*klimata*) of the Earth. Each κλίμα (*klima*) is first defined by the length of its longest day, which is given a specific value. The varying lengths of daylight in the different κλίματα form an arithmetic progression.[38]

The word κλίμα (or ἔγκλιμα) is geometric in character, meaning "inclination" or "slope." As Neugebauer explained, in the astronomical context it means "the inclination of the earth's axis with respect to the plane of the local horizon." He pointedly remarked that "this concept presupposes the discovery of the sphericity of the earth." The *klimata* were parallel latitudinal bands of the Earth which distinguished the different inclinations of the planes of the various horizons towards the Earth's axis. Neugebauer emphasized the mathematical meaning of the word *klima*, noting that it is "of fundamental importance for the understanding of the historical development of the concept 'climata' to realize that it has its origin in problems of spherical astronomy."[39]

Neugebauer's insistence on the primacy of the astronomical application of *klimata* is supported by a statement from the geographer Strabo, who had recognized the importance of the *klimata* for astronomy, for in his own treatment of the subject he claimed that astronomers must consider it more fully. He believed that geographers must be dependent on the work of astronomers, and credited Hipparchus with having stated in his *Against Eratosthenes* that

> for everyone, both laypeople and scholars, it is impossible to obtain proper knowledge of geographical description without the determination of the heaven and the observed eclipses. It is not possible to know whether Alexandria in Egypt is further north of Babylon or south, or how much the distance is, without investigation of these things by means of the *klimata*.

Neugebauer argued that the use by geographers of the *klimata* "as an ordering principle of empirical geographical material—lists of cities, ethnographic characteristics of zones, etc.—represents a secondary development." Ptolemy himself differentiated between astronomical and geographical considerations at the end of Book Two of the *Syntaxis* by stating that the discussion of the details of the *klimata* with regard to the distance of cities from the equator and from Alexandria belong in a separate treatise on geography, a work which he later wrote.[40]

However, Strabo pointed to the common ground of astronomers and geographers when he noted that "it is one of the things proper to geography to hypothesize (ὑποθέσθαι) that the Earth as a whole (τὴν γῆν ὅλην) is sphere-shaped."[41] It is following this statement that Strabo then discussed at length the various geographical *klimata* or zones (ζῶναι).

Nevertheless, Strabo's stated reasons for adopting this idea are rather different from those given by Ptolemy. Strabo wrote that the notion of the Earth's sphericity comes both from ideas (for example from the idea of the motion toward the center from farther away and from the idea of each body inclining toward its own place of suspension) and also from the appearances (for example those of the sea and the heaven). Both the ideas and the appearances are confirmed by the senses and also by common intuition.[42] Strabo seems to have accepted both the arguments of Aristotle and the common phenomena, whereas Ptolemy relied primarily on common experience.

But of course Ptolemy worked not only within the Greek astronomical tradition, but within the geographical tradition as well. His work, which included the *Geography* as well as the *Tetrabiblos,* shows his participation in the latter tradition.

When he mentioned that lunar eclipses, which take place simultaneously for all observers, are not recorded as taking place at the same time by those in the East and the West, Ptolemy was implicitly referring to the idea of geographical longitude. Geographical longitudes, and the usefulness of differences in local time for their determination, were, like the assignment of *klimata*, based on the theory of the spherical Earth. À propos of this is Strabo's emphasis on the importance of astronomy for the study of geography, and his crediting Hipparchus for having realized this importance.[43]

Strabo suggested that Hipparchus used information about of both lunar and solar eclipses to determine local time differences; nevertheless, solar eclipses are valueless for the determination of longitude. A lunar eclipse occurs when the Moon enters the Earth's shadow, while a solar eclipse occurs when the Moon intersects the line of vision to the Sun of an observer on Earth. Because the shadow of the Earth on the Moon is not influenced by the place of the observer on Earth, all observers see all stages of a lunar eclipse simultaneously. In the case of a solar eclipse, however, the place of the observer on Earth does influence the phenomenon, which is dependent on the observer's line of sight to the Sun (lunar parallax must be taken into account).[44]

Ptolemy was aware of the special significance of lunar eclipses for the determination of longitude and the differences in local time, for in chapter one of Book Two, he noted that the same eclipses, especially lunar eclipses, observed both in the extreme west and the extreme east are not earlier or later by more than twelve equinoctial hours. Although, as Neugebauer has noted, at least in principle "it should have been easy enough to establish reliable data since lunar eclipses are very frequent and their occurrence easily predictable," this does not mean that Ptolemy (or anyone else, for that matter) had records of observations of lunar eclipses at their disposal. Toomer sensibly offered the cautionary remark that it should not be inferred from this "that Ptolemy possessed records of lunar eclipses observed simultaneously at eastern and western ends of the known world."[45]

Both Toomer and Neugebauer agreed that Ptolemy probably had a record of only one lunar eclipse observed in two places, that observed at Arbela and Carthage on September 20, 330 B.C. (famous because of the battle at Arbela). Furthermore, Neugebauer has noted that the data for this eclipse

> were also only approximately known, so that Pliny and Ptolemy give widely different data. Interestingly enough Pliny's data are superior to Ptolemy's, both with respect to the time interval and to the phases of the eclipse.

However poor the records of simultaneously observed lunar eclipses available to Ptolemy may have been, he was nevertheless aware of their theoretical value in the determination of geographical longitude.[46]

III. *That the Earth is in the middle of the heavens, with regard to the senses*[47]

To support his hypothesis that the Earth is in the middle of the heavens, Ptolemy relied on arguments based on astronomical observations, rather than considerations of any physical theory. In his justification of the hypothesis stating the central position of the Earth, with regard to the senses, he noted that if the Earth were not in the middle of the heavens it could only be situated in a position: (1) not on the axis of the universe but equidistant from each of the poles; or (2) on the axis but displaced toward one of the poles; or (3) neither on the axis nor equidistant from the poles.

He provided the following arguments against the first possibility. First, if the Earth were either above or below the axis of the universe, equinoxes either would not occur at all or would not occur when they do, because the horizon would bisect not the celestial equator, but one of the circles parallel to it. Ptolemy pointed out that this cannot be the case, since "everyone agrees" that the equinoxes occur at the same time everywhere, that is, midway between summer and winter solstice.

If the Earth were moved to the east or west of the axis, "the sizes and distances of the stars would not appear equal and the same at the eastern and western horizons." Here, Ptolemy was presuming that the displacement of the Earth with regard to the axis must be large relative to the size of the heaven. He had a further objection to this possible position of the Earth, because the consequence would be that the sizes and distances of the stars would not appear to be equal at the eastern and western horizons, nor would the time for rising to the middle of the heaven be equal to the time for setting from the middle of the heaven. Neither of these occurrences would be in accord with the phenomena. If the Earth were on the celestial axis, but closer to one pole than the other, other anomalies would result. The horizon would divide the heavens into two unequal parts, as well as the ecliptic ("the great circle through the middle of the zodiac"). However, since at all times and places six zodiacal signs are visible above the Earth, while six are invisible, the horizon must bisect the ecliptic and the heavens. Further, on the day of equinox, the

Figure 3.10 The stars appear to be at the same distance from the Earth ([*Sphaera mundi*], [Venice], Erhard Ratdolt, 1485, [p.8]).

Figure 3.11 If the Earth were not in the middle of the heavens, the horizon would not bisect the zodiac (Sacro Bosco, *Sphaera,* Lyons, 1564, p. 23).

shadow cast by a gnomon at sunrise would not form a straight line with the shadow cast at sunset, which would be contrary to what is observed.

Ptolemy rejected the third possible position of the Earth by referring to the above arguments, which would also apply in this case. He added that if the Earth were not in the center of the universe, eclipses of the Moon would not occur only when the Moon is in a position diametrically opposed to the Sun, since the Earth would often come between them when they were not diametrically opposed to one another, but at intervals of less than a semicircle.[48]

The mathematical and astronomical consequences of the Earth's shape and position seem to have been most important for Ptolemy. Physical considerations and arguments had no place in Ptolemy's justification of the idea that the Earth is at the center of the universe. While Aristotle would have agreed with Ptolemy that the Earth is at the center of the cosmos, Aristotle offered arguments of an entirely different type to support the same thesis.

In his arguments in support of the central position of the Earth, Aristotle again used physical arguments which rely on his doctrine of natural motion. In Book One of *On the Heavens,* in order to prevent the possibility of other worlds, Aristotle asserted that there can be only one center of the universe. Aristotle argued that even if several worlds did exist, these worlds must be composed of the same elements with the same natural motions. All portions of the element earth must be located by nature in one place, the center of the universe. In Book Two, within the context of a larger discussion concerning whether the Earth is in motion, Aristotle discussed the position of

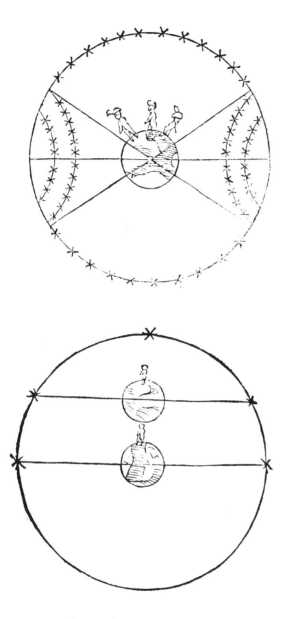

Figure 3.12 The Earth appears to be in the middle of the heavens (Francesco Barozzi, *Cosmographia in qvatvor libros,* Venice, 1585, pp. 24, 30).

the Earth, explaining that the motion of the parts having weight and of the whole Earth itself is naturally toward the center of the universe. For this reason the Earth happens to lie at the center. According to Aristotle, it is natural for the whole Earth, as well as parts of the Earth, (that is, earthy, or heavy things) to move toward the center of the universe.[49]

Ptolemy's neglect of Aristotle's argument should not be misinterpreted as evidence that the physical considerations were of no importance in Ptolemy's time. There is evidence that the physical issues were still alive in the first and second centuries A.D. For example, although Epicurean cosmology incorporated the idea of the central position of the Earth, the Epicureans rejected Aristotle's explanation of motion in terms of natural place and, as a consequence of their own explanation of motion, rejected the idea of the sphericity of the Earth in favor of a flat Earth. Lucretius explained that the Earth is supported in the center of the universe; it is not a burden and does not weigh down the air. This is so because the Earth is a living part of the universe; it is not a burden to the air, just as to a man his limbs are not a burden, and the head is not a burden to the neck. But even though Lucretius accepted the idea of the central position of the Earth, he rejected Aristotelian physics. He cautioned against the acceptance of the idea that motion is toward the center of the universe, stating that with regard to motion, there is no privileged place in the universe.[50]

Lucretius's warning may have been directed against the ideas of the Stoic philosophers. According to Stobaeus, Zeno of Citium had said that the Earth is in the middle of the *kosmos,* but the reasons given to support this view are not Aristotelian:

> all the parts of the cosmos have a motion toward the middle of the cosmos, especially those having weight. This same cause serves both for the stability of the cosmos in an infinite void and likewise of the Earth in the cosmos, being situated in equilibrium at the center.

Zeno's argument for the central position of the Earth is in some ways reminiscent of Aristotle's, in that it depends on motion toward the center. However, Zeno deviated significantly from Aristotle's physics, by stating that air and fire, even though they are weightless, also move toward the center. Furthermore, for Zeno, the Earth is not only in the middle of the *kosmos,* the *kosmos* is in the center of the infinite void.[51]

In the dialogue *On the Face of the Moon,* written by Plutarch, the speakers discuss the position of the Earth. Pharnaces, one of the participants in the dialogue, relied on physical considerations and an argument which is similar to Aristotle's own to support his position that the Earth is in the center of the cosmos. His position was attacked by Lamprias, who listed a series of objections to the idea that the Earth is spherical, some of which were based on observation. He then attacked the idea of the central position of the Earth, and considered some rather bizarre consequences of heavy things converging on a single point.[52]

Plutarch wrote a diatribe against Aristotelian, and also presumably Stoic, physics, which he placed in the mouth of Lamprias. Rejecting the doctrine of natural motion, he offered his own views and suggested that the Earth acts as a center of motion toward which only some things move; it is not the only center in the universe toward which there is motion. Celestial bodies, including the Sun and the Moon, may also serve as centers of motion and different elements might have different centers of motion.[53] He made this suggestion as part of his objection that it did not make sense that all heavy things should tend to converge in one place, if light things do not as well. Lamprias carried the Aristotelian concept of natural motion to an extreme in his argument against the idea that the Earth is the center of the universe. Clearly, Plutarch, in writing this dialogue, conceived of the possibility of many centers of attraction existing in the universe.

It is crucial to remember that Aristotle's arguments in favor of the idea of the central position of the Earth were part of his physical theory, particularly linked to the doctrine of natural place and motion. Ptolemy's arguments were not physical; they did not touch on the problems involved in accounting for the motion of objects toward Earth. While his arguments did discuss the motion of celestial bodies, he was, in discussing the central position of the Earth, only concerned with accounting for how the motions appear to observers, not in explaining how the motions occur. His stated reasons for accepting the idea of the centrality of the Earth are not Aristotelian; neither are they Epicurean or Stoic. Simply stated, they are not motivated by any natural philosophy. There were many physical arguments, not only those of Aristotle, available to support the idea that the Earth is in the center of the universe. Ptolemy used none of them, relying instead on arguments based on observation of the heavens which supported the tenets of spherical astronomy.

However, such arguments do not depend on the Earth being mathematically in the center of the universe, as Ptolemy realized. In the first sentence of his discussion of this hypothesis, he stated that the Earth is in the center of the universe, with regard to the senses. Ptolemy clearly recognized that, with regard to the tenets of spherical astronomy, the Earth's position in the center of the universe need be only approximate. Furthermore, there is the suggestion that while the Earth may be perceived by the senses as being in the center, mathematically it may not be. To fulfill his self-appointed task of justifying his hypotheses by their agreement with the phenomena, Ptolemy needed only to show that the appearances which are perceived do agree with his hypothesis. Without a basis in Aristotle's theory of natural place, the argument for the strict centrality of the Earth is weak. Ptolemy, however, did not argue the strict centrality of the Earth, recognizing that the perception of centrality was all that was required for his purposes.

IV. *That the Earth has the ratio of a point relative to the size of the heavens*[54]

Ptolemy's fourth physical hypothesis is that the Earth has the ratio of a point with regard to the size of the heavens. Aristotle never discussed this idea, but only hinted at it. As part of his larger argument concerning the sphericity of the Earth, Aristotle stated that the size of the Earth is small.[55]

Aristotle offered evidence based on astronomical observation to support his statement. He noted that even a small change in an observer's position northward or southward results in a different view of the stars. He pointed out that certain stars are visible in southern regions which are not seen in the north, and that some stars remain visible in the north which are seen to set in the south. He stated that "thus not only is it clear from these considerations that the shape of the Earth is spherical, but also a sphere which is not large." Otherwise, such a small change of position on the part of the observer could not have had such an effect.[56]

He then offered additional evidence for the small size of the Earth. He stated that "those who imagine that the region around the Pillars of Hercules reaches to that around India, and that this way the ocean is one, would seem not to be imagining something unbelievable," noting that such people offer in support of their suggestion the fact that elephants are found at the extremities of both lands, and arguing that this is due to the connection between the two. While Aristotle's suggestion that travel may be accomplished fairly easily may be an argument for the small size of the Earth, it does not directly address the question of the size of the Earth relative to the universe as a whole.

Like Aristotle, Ptolemy also considered the size of the Earth to be relatively small; however, he was particularly interested in its size in relation to its distance from the sphere of the fixed stars, that is, the size of the entire universe. He stated this quite clearly, claiming that "the Earth has sensibly the ratio of a point to its distance from the sphere of the so-called unwandering stars."

The first piece of evidence offered by Ptolemy, like that offered by Aristotle to support the claim of the small size of the Earth, was observational:

> from all parts of the Earth everywhere, the sizes and distances of the stars, at the same time, appear equal and the same, just as the observations of the same [celestial] objects from different *klimata* are found to have not the least disagreement.

Both Aristotle and Ptolemy referred to what an observer sees when he changes his position. For Aristotle it was significant that the view seems to change; different stars are seen in the north and in the south. While Ptolemy was certainly aware of this, and its importance for determining *klimata*, he emphasized something quite different. He stressed the similarity of what is seen; that is, the stellar positions remain the same, relative to one another.

As another piece of evidence to support the hypothesis that the size of the Earth is relatively small, Ptolemy added that gnomons and the center of a ringed sphere could not serve mathematically as the center of the Earth unless the Earth were very far away from the celestial bodies. This justification is based no doubt on his familiarity with two of the most important instruments available to ancient astronomers. However, the development and use of these instruments was in large part based on the theory of the sphericity of the Earth and the heavens, namely, Ptolemy's first two hypotheses.

The assumption of yet another of his hypotheses is apparent also in the final piece of evidence Ptolemy offered to justify the hypothesis that the Earth has the ratio of a point to the heavens. He stated that the horizon always bisects the sphere of the heavens, recalling his argument that the Earth is in the center of the universe:

> the planes drawn through the eyes, which we call 'horizons', always cut the whole sphere of the heaven in two, which would not happen if the size of the Earth was perceptible with regard to the distance of the heaven, but in that case only the plane through the point at the center of the Earth would be able to bisect the sphere.[57]

In other words, the entire Earth as a whole must be small enough to function as the center-point of the universe.

The actual size of the Earth is not at issue here; rather, what is important is the size of the Earth relative to the universe. The size of the Earth is so large that its sphericity is not immediately perceptible, but must be inferred. In spite of this large size (relative to human observers), the Earth is small relative to the heaven, small enough so that it can function, observationally, as the center of the universe even if it is not geometrically the center. The arguments offered by Ptolemy to support his third hypothesis, that the Earth is in the middle of the heaven, support the view that the Earth is, on the basis of observational data, in the center.

That, in the *Syntaxis,* Ptolemy considered perceptible parallax to be the key to the determination of the distances of celestial bodies is shown by his criticism of the methods used by Hipparchus in determining the distances of the Sun and Moon. He criticized Hipparchus, noting that he attempted to demonstrate the Moon's distance by guessing at the Sun's distance, that he sometimes assumed that the Sun displays no perceptible parallax while at other times he assumed that it did display parallax. Ptolemy pointed out that not only the size of the Sun's parallax, but whether it displays parallax at all is at issue.[58]

When Ptolemy stated his hypothesis that the Earth has the ratio of a point with regard to the distance of the sphere of the fixed stars, he was, in essence, stating that the size of the universe is very large. Ptolemy needed a large universe to account for the absence of observed stellar parallaxes from different points on the Earth. Ptolemy knew that theoretically the distances of the celestial bodies could be determined if parallax could be observed, using the same method as he applied to lunar parallax. Concerning those celestial bodies for which no parallax was observed, he reasoned that they must therefore be very much farther away from the Earth than the Moon. He described the celestial bodies with no perceptible parallax as "those to which the Earth has the ratio of a point."[59]

When Ptolemy claimed that the Earth has the ratio of a point relative to its distance from the fixed stars, he was following the tradition of mathematicians. In the first proposition of the *Phenomena,* Euclid stated that the Earth is in the middle of the universe, and occupies the place of center in relation to the universe. This was an idea which gained a following among mathematicians. Aristarchus had stated that "the Earth is in the relation of a point and center to the sphere of the Moon." Geminus said that the whole Earth has the ratio of a center to the sphere of the fixed stars. Ptolemy's concern with the lack of perceptible parallax indicates that observational concerns as well as mathematical ideas motivated his adoption of the hypothesis.⁶⁰

V. *That the Earth makes no motion involving change of place*⁶¹

Of the hypotheses so far considered, it is Ptolemy's justification of this one, maintaining that the Earth does not move from place to place, which is most natural-philosophical in character and which, more than any others of the hypotheses, shows a debt to Aristotle both in terms of ideas and method of argument. However, Ptolemy was not, even here, a strict Aristotelian. In the course of his discussion of this hypothesis, Ptolemy also considered some other physical issues, such as the natural motion of heavy bodies toward the center, and the means by which the Earth maintains its position in the center of the universe.

Ptolemy began his discussion of this hypothesis by stating that the same arguments which support the view that the Earth occupies the center of the universe also support his belief that the Earth does not experience motion from place to place. He stated that the same phenomena would result if the Earth were in motion as those phenomena that would be apparent if the Earth had any position other than the center. It is striking that in this chapter, which is the longest of those dealing with his physical hypotheses, this is Ptolemy's only justification based on astronomical considerations. It is a very brief allusion, at that. The other arguments in this chapter are all based on physical considerations. In this, he seems to have been influenced by Aristotle's own discussion of the immobility of the Earth.

Aristotle had listed several objections to the idea of the motion of the Earth in *On the Heavens*. His first argument is based on his general theory of motion. He noted that some people make the Earth "one of the stars, but others, putting it at the center, say that it is winding and moving about the middle pole." He objected to these possibilities, and presented the following argument in Book Two to support his view that the Earth, as a whole, is at rest in the center of the universe.[62]

Aristotle based his argument concerning the lack of motion of the entire Earth on observations of the motion of pieces or parts of Earth and his own philosophical views concerning natural motion. He stated that all pieces of Earth move toward the center in a straight line, by nature. This natural motion is part of the order of the universe and is unchanging. Aristotle stated that if the Earth as a whole moved, "either at the center or away from the center, its motion must necessarily be by force," it would not be the natural motion of Earth. If the Earth as a whole moved away from the center this motion would be violent motion since "it is not the motion of the Earth itself." For the natural motion of Earth is rectilinear towards the center of the universe, as Aristotle had already clearly stated. Any parts of Earth which do move away from the Earth are moving by force, only temporarily. Having already established that the natural motion of Earth is toward the center of the universe, Aristotle objected to any motion of the Earth as a whole which was not linear toward the center of the universe as being unnatural and contrary to the eternal order (τάξις) of the cosmos, an objection which was based on his need for internal consistency within his own theoretical framework.⁶³

If the Earth as a whole moved "at the center," this motion too would be unnatural. However, Aristotle did not state this objection directly. Aristotle might have reasonably made the objection that axial motion at the center of the universe by the Earth as a whole would be impossible since this would involve a circular motion, which is not the natural motion of Earth, and such a circular motion could only be temporary. He did not, however, make this argument, instead suggesting that if the Earth as a whole moved at the center of the universe, this motion must, by analogy to those bodies which experience circular motion, have a double motion (presumably, that of the first sphere and that of the body itself). Aristotle stated that

> all the moving bodies manifest a circular motion which lags behind and are moving with more than one motion, except for the first sphere. It is necessary for the Earth as well to have two motions, whether moving around the center or remaining at it.

But, he explained, if this were so, the stars would not appear as they do, rising and setting at the same places, but would exhibit other motions. And, if the Earth had such a double motion, the celestial phenomena would not appear as they do.⁶⁴

This second objection to the idea that the Earth is in motion is based on the appearances of the heavens. Here, Aristotle argued that if the

Earth moves in any way, this motion must be similar to that of the heavenly bodies. By analogy, the Earth would also have a double motion (a daily motion as well as an annual motion against the background of the fixed stars). But, said Aristotle, the appearances do not support a double motion of the Earth, so therefore it cannot move. Thus, Aristotle used two kinds of arguments to support the idea that the Earth is in the center of the universe, one based on his doctrine of natural motion, the other based on observational evidence.

Aristotle brought up the evidence of celestial phenomena as an objection to the idea that the Earth is in motion; so did Ptolemy. However, Ptolemy's own objection that the appearances do not support the idea of the Earth being in motion was very briefly stated; he essentially directed the reader to his arguments in support of the hypothesis that the Earth is as a point in relation to the universe, and did not draw out his argument in the style of Aristotle.

Aristotle's third objection to the idea of the Earth experiencing motion is, like his first objection, also grounded in his theory of natural motion. He began by explaining why the Earth is at the center of the universe, stating that because the natural motion of earthy bodies is toward the center of the universe, the Earth as a whole is at the center, incidentally. He also noted that there might be some question about whether the natural motion of heavy bodies, as earthy bodies, is towards the center of the Earth or toward the center of the universe, since the center of both is the same point. He explained that it must be toward the center of the universe, because light bodies move to the region away from the center.

Here Aristotle considered, albeit rather briefly, the reasons why heavy bodies fall toward the center of the Earth. It is only incidental that heavy bodies move toward the center of the Earth; their motion is best understood as being toward the center of the universe. The center of Earth should be understood as a remote final cause, for in Book Four Aristotle explained that if the Earth were moved to where the Moon is now, each of the parts of the Earth would not move toward the whole Earth, but to the place where the Earth is now, that is, the center of the universe. Aristotle then provided evidence that heavy bodies move toward the center of the Earth: "moving heavy bodies do not move in parallel lines but at similar angles, thus they move to the center, indeed, the center of the Earth." The geometrical character of this description is somewhat uncharacteristic.[65]

It is striking that Ptolemy, immediately after briefly considering the evidence based on astronomical observation which supports the idea that the Earth does not move from place to place, discussed essentially the same points as Aristotle, in the same order of presentation. But he did not touch on every point mentioned by Aristotle. The points of similarity suggest that Ptolemy used *On the Heavens* as he was compiling his reasons to support this hypothesis. Ptolemy, nevertheless, did not repeat Aristotle's arguments verbatim, and presented his own point of view.

In his own discussion of the immobility of the Earth, Ptolemy alluded to Aristotle's theory of motion and seems to have generally accepted it. He began his discussion by noting that the arguments which he has already presented also argue against any motion of the Earth. Ptolemy objected to the idea that the Earth is in motion, because such an hypothesis would result in the same difficulties as would arise if the Earth were not sensibly in the center of the universe. He claimed that either hypothesis would require the observation of certain phenomena which are not actually observed.

Having stated his initial objection to the idea that the Earth is in motion, Ptolemy stated that

> it seems to me that it would be useless to look for any causes for the motion toward the center, once it is clear from the phenomena themselves that the Earth is in the center of the universe, and that all heavy objects move toward it.

Although Ptolemy did not mention Aristotle by name, in the context of his rejection of the search for final causes, it seems possible that he had Aristotle in mind as he wrote this passage.

Like Aristotle, Ptolemy considered the evidence of astronomical observation and then turned to the topic of motion toward the center of the Earth. But Ptolemy treated both very briefly. He then brought up the same evidence as Aristotle had to support the idea that heavy objects fall toward the center of the Earth, stating that both the directions and paths of proper ($\iota\delta\iota\alpha\varsigma$) motion of those bodies having weight is always and everywhere at right angles to the rigid plane drawn tangent to the impact.[66] He concluded by noting that "it is clear because of this being so that, if not stopped by the surface of the Earth, they would certainly fall to the actual center, since the straight line leading to the center is always at right angles to the plane tangent to the sphere," at the point with the intersecting line. It is noteworthy that Ptolemy's geometrical description of the paths of heavy bodies falling toward the center of the Earth does not merely echo Aristotle's own, but is more precise.

Up to this point, Ptolemy was following Aristotle's line of argument very closely, but stated it in a much briefer form. Here, as soon as something of mathematical interest appears, Ptolemy elaborated on Aristotle's theme. So far, Ptolemy's principal argument against the Earth's motion from place to place was based on the idea that the Earth must be at the center of the universe for largely physical reasons, because of the motion of heavy bodies. However, it is at this point that Ptolemy began to deviate from the order of Aristotle's argument, which continued as follows.

Aristotle very briefly mentioned, apparently as an afterthought, that "heavy objects being thrown upwards by force in a straight line again come back to the same place, even if the force casts them to an unlimited [distance]."[67] While this does not directly address the question of the Earth's immobility, it is evidence that earthy bodies fall to the Earth and fits into his larger argument concerning the position of the Earth at the center of the universe. Although up to now Ptolemy followed Aristotle's order of argument closely, he began to deviate here and did not mention this point.

Aristotle concluded his objections to the idea of the motion of the Earth by reiterating his argument that his physical explanation of motion is sufficient to demonstrate that the Earth does not move from place to place, arguing that since earth by nature moves from everywhere to the center, it is impossible for any earthy portion to move away from the center of the universe, except by force. He concluded his argument by noting that if "any particular portion is not able to move away from the center, it is manifest that the entirety is still more incapable." For the whole naturally has the same motion as a portion of the whole. Since it cannot move except by the agency of a stronger force, it necessarily must remain at the center. Aristotle received his support for the idea of the Earth's immobility from considerations based upon his own physical theory of motion.[68]

Following his discussion of the physical reasons for the Earth's immobility, Aristotle noted that this idea is confirmed by what the mathematicians have to say about astronomy, for the phenomena are consistent with the notion that the Earth is at the center.[69] Once again, Aristotle used observation to corroborate physical theory. Aristotle had made it clear that the physical considerations are sufficient to justify why the Earth must be immobile; that conclusions drawn from astronomical observations agree with his conclusions drawn from his own theory of motion is convenient, but not sufficient to explain the immobility of the Earth.

Having already deviated from Aristotle's line of argument, Ptolemy next confronted the problem of accounting for how the Earth remains unsupported and unmoving in the center of the universe. As part of the larger discussion, Ptolemy referred in passing to the question of whether or not there is directionality with regard to the universe as a whole. His comments are worth examining because of the philosophical allegiances they disclose. While they were in agreement that the Earth is in the center of the universe, Plato and Aristotle disagreed as to whether direction terms, such as "up" and "down" should be applied to describe the universe as a whole.

In Plato's dialogue of the same name, Timaeus explained that the "heavy" and the "light" can best be understood when examined in connection with the notions of above and below.[70] This statement provides an important clue as to why Ptolemy included his discussion of directionality in the universe within his larger discussion of how the Earth maintains its central position. Ptolemy agreed with Plato that the understanding of the "heavy" and the "light" can be aided by a consideration of directionality.

In order to understand Ptolemy's views on directionality in the universe, it will be necessary to examine those of Plato and of Aristotle, who clearly disagreed with Plato on this subject. Timaeus stated that it is a mistaken notion that the universe is divided into two regions, a lower one toward which all things with weight tend to move, and an upper one to which things only move involuntarily. Timaeus employed a geometrical argument against using direction terminology to describe the universe, explaining that since the whole heaven is spherical, all the extremities are equidistant from the center and are therefore equally extreme from the center. The center, which is equidistant from them all, must be regarded as the opposite to all of the extremities. Since this is the nature of the world, anyone describing these extremes as above or below is not speaking properly. Furthermore, no part of the Earth, which is in the center of the universe, should be described as "above" or "below," but rather simply as at the center. Timaeus concluded this argument and then went on to account for the common terms used to describe directionality in the world. Following this, Timaeus went on to discuss the relative nature of the terms "heavy" and "light;" the discussion of direction terminology was used by Timaeus to facilitate the consideration of this larger issue.[71]

Whether or not one may rightly speak of direction in the universe was a question which Aristotle felt still required resolution. He referred to the directions of natural motion in many places. In the *Physics,* he briefly described what he meant by directions in the universe,[72] but explained the use of direction terminology most fully in *On the Heavens.* His discussion was probably aimed at the *Timaeus,* for it is also part of a larger discussion of heaviness and lightness. Aristotle paraphrased Plato at certain points and also, like Plato, commented on popular usage.

Aristotle first commented on his own use of direction terminology in his chapter on heaviness and lightness, noting that "there are some things which naturally move away from the center, and other things which always move toward the center." He explained that he described those things which move away from the center as moving upwards, and those which move toward the center as moving downwards. He noted that "some people" (implying they would themselves know who they were) claim that there is no up or down in the world, and provided a statement of their position (a position which he clearly rejected, since he pointedly contrasted it with his own): "there is no up or down, they say, since [the universe] is the same everywhere and someone walking would in each place be at his own antipodes."[73]

This last statement appears to be a paraphrase from the *Timaeus,* for Timaeus stated that "if someone would walk around [an hypothetical body at the center of the universe] in a circle, he would often be standing at [what had previously been] his own antipodes and [would then, at different times] speak of the same place as above and below," a situation which Timaeus regarded as nonsensical. Timaeus had explained that because the whole is spherical, it makes no sense to speak of one part as up and the other as below.[74] An element of ridicule present in the language of both Plato and Aristotle suggests that this was a sore spot of disagreement between them.

Aristotle went on to clarify his own terminology, explaining that he applied "up" to the extremity of the universe, because it is "both upward in position and primary by nature." He argued that "since the universe has both an extremity and a center, it is clear that there must be an up and down," and called upon common usage as support for his terminology, stating that "this is just as οἱ πολλοὶ [most people] say, except they don't speak precisely." This is because the average man does not realize that the hemisphere above is only one half of an entire sphere; if he did, according to Aristotle, he would call the extremity up and the center down.[75] Clearly, Aristotle thought his use of direction terminology was very reasonable.

Ptolemy's own discussion of direction terminology shares with those of both Plato and Aristotle an interest in accounting for common usage. However, Ptolemy clearly rejected Aristotle's opinion that directionality exists, in favor of a more geometrical treatment, reminiscent of Plato's. Ptolemy stated that "there is no up and down in the cosmos with respect to itself, just as one would not imagine such a thing in a sphere."

He then explained common direction terminology. Light and rarefied things move toward the outside and the circumference, and seem to move upward only because we call the direction above our heads "up"; heavy and dense bodies move toward the middle and the center, but seem to fall downward only because we call the direction of our feet "down." While Ptolemy agreed with Plato's point of view, that the common terminology is grossly misleading, he offered a much simpler, and completely different, explanation of common usage than did Plato.

Plato's explanation is lengthy and complicated, and will only be briefly summarized here. In order to explain direction terminology, Timaeus used a thought experiment. His argument was as follows: if we imagine a man standing in that part of the universe which is reserved for fire, and breaking off different-sized chunks of fire to weigh them, he will have an easier time managing the smaller chunk, because smaller things are handled by force more easily than bigger things. In this case, even though the bigger chunk of fire should be called light, it will be "heavy" and will be described as moving downwards, because it is harder to manage than the smaller piece of fire. Timaeus realized that this is just what happens in our own region, that what is "heavy" and "light" is a relative distinction which varies, depending on the nature of the body itself and its environment. Because of the relative nature of direction terminology, Timaeus complained about the arbitrariness of this system of terminology.[76]

Timaeus had begun his objections to the use of direction terminology to describe the universe by invoking geometrical considerations; he ended with an argument based on the physical problems of heaviness and lightness. Ptolemy rejected this explanation of direction terminology in favor of a more straightforward one based not on physical considerations, but instead on the geometry of the sphere, providing an explanation to appeal to the mathematician in all of us.

Even though he accepted Plato's, rather than Aristotle's, position on whether directions exist in the universe as a whole, and whether direction terminology may be properly used, Ptolemy generally opted for Aristotle's physical theory and his methodological approach in his own discussion of how the Earth remains in the center of the universe, of which the discussion concerning directionality was only a part. But Ptolemy did differ from Aristotle somewhat in his physical explanations.

Whereas Ptolemy often stressed the importance of observational evidence, here he did an about-face, discounting experiential evidence in favor of theoretical considerations, stating that

> those who think it is a paradox that so great a weight as the Earth is neither somehow supported nor moves seem to me to be making a mistake by looking at their own experience and not the peculiarity of the universe.

Relying on his previous hypothesis regarding the relatively small size of the Earth, he suggested that those who object to the idea that the Earth remains motionless in the center of the universe without support would not find it so implausible if they understood that the "size of the Earth being compared to the whole body surrounding it has the ratio of a point to it." He argued that it is quite possible that the Earth, which is relatively smallest, should be overpowered and pressed against equally from all sides by that which is the greatest of all and of uniform nature, namely the universe as a whole.

It is here that Ptolemy discussed whether there are directions in the universe. Following this, he then explained that heavy bodies settle around the center because of mutual resistance and pressure, which is equal and uniform from all sides. He concluded that since the whole Earth is so much larger than these relatively small weights falling against it, it is unmoved by the assault of these small weights.

Aristotle presented a rather lively discussion of the problem of accounting for how the Earth keeps its place at the center of the universe in *On the Heavens*. He noted (294a19–22) that those who have discussed the motion and rest of the Earth have done so in various ways and that it is certainly a puzzling issue about which almost everyone has wondered. This has consequently become a general topic of study. He expressed astonishment that the solutions offered have not been seen as more bizarre than the problem itself.

After criticizing the ideas of his predecessors, Aristotle gave his own answer to the problem of how the Earth maintains its position at the center of the universe. Earlier in *On the Heavens,* he had asserted that no substance has more than one natural motion. Since earth and other heavy substances fall towards the center, that is their natural motion; they can have no other (269a30–269b3). Because the natural motion of earth is toward the center of the universe, the Earth is at the center. It is "impossible for any portion of earth to move from the center except by force." Since the Earth "cannot move except by means of a more powerful force, necessarily it must remain at the center."⁷⁷

In his arguments against the notion of the Earth being in motion, Ptolemy did not refer to the doctrine of natural place. He very briefly stated that it cannot be the case that the Earth has a single motion in common with other heavy objects, because if it were so "living things and heavy objects would be left carried in the air," and the Earth would quickly fall completely from the heaven. He concluded by remarking that it is laughable (γελοιότατα) to think about such things.

Ptolemy then confronted the suggestions of others (perhaps Heraclides of Pontus and Aristarchus) that the Earth is in motion. Without naming any names, he stated that certain people have suggested that there is nothing to oppose the view that the heaven is motionless and the Earth revolves on an axis from west to east, or, alternatively, that both heaven and Earth move about the same axis in a way which preserves their overtaking of one another. He admitted that there is nothing in the appearances of the stars which would contradict such ideas, but he objected to such theories on the basis of physical phenomena which occur in the air, in the terrestrial region, scorning such views as ridiculous. He invoked physical theory to argue against the possibility of the Earth's motion.

Ptolemy suggested that, for the sake of argument, it be imagined that the aether (which is the lightest and most rarified of matter) either has no motion, or else has a motion no different from that of matter which is of an opposite nature (that is, heavy and dense). He was also willing to imagine that heavy objects could have a rapid motion naturally, even though he pointed out that these objects are sometimes very difficult to move even by force. He argued that even if one is willing to imagine such things, nevertheless, those who propound a theory of the Earth's motion must admit that if the Earth made a revolving motion it would be the most violent of all motions because it would be accomplished in so short a time. He reasoned that the result would be that all objects not always standing on the Earth would appear to have a motion opposite to that of the Earth: "neither clouds nor other objects flying or being thrown would ever be seen moving toward the east." The Earth's motion toward the east would always overtake such objects, so that they, being left behind, would seem to move toward the west. Ptolemy argued that even if one admits ideas which seem ridiculous on the basis of physical theory, one must conclude that the appearances argue against the motion of the Earth.

He also objected to the idea that the air is carried along with the Earth, noting that if the air is carried along in the same direction and speed as the Earth, one of two things would occur. Either the compound objects in the air would always appear to be left behind by the motion of both the air and the Earth, or, if the compound objects were also carried around in the air they would then never appear to have any other motion and would appear to be at rest; flying or thrown objects would not appear to be in motion. He dismissed this last possibility as being plainly in contradiction with experience. Ptolemy based his final objection to the motion of the Earth on the evidence of everyday observation.

In his justification of the hypothesis that the

Earth has no motion, Ptolemy relied, to some extent, on the pattern of Aristotle's argument against the idea of the Earth's motion. He did not copy Aristotle's argument, but, rather, summarized some points, while elaborating on others. He deviated from Aristotle in his position that there is no directionality in the universe, instead adopting a point of view similar to that presented in Plato's *Timaeus*. Here he did not copy Plato's argument, but presented his own, much simpler, account of the common usage of direction terminology to describe the universe. Finally, he confronted and argued against theories of the Earth's motion by appealing to commonplace empirical evidence.

VI. *That there are two different primary motions in the heavens*[78]

Ptolemy's final hypothesis is that there are two different primary motions in the heavens: (1) the daily motion, which "carries everything from east to west," and (2) the motion of the Sun, Moon, and five planets (roughly) along the ecliptic from west to east. The first motion he described as follows: "it carries everything from east to west always in the same manner and at the same speed, making the revolutions along circles parallel to each other, being clearly described about the poles of this sphere which rotates everything uniformly." He described the second motion as being that by which "the spheres of the stars make an opposite motion to the first motion, around another pair of poles, which are not those of the first rotation."

The essential characteristics of Ptolemy's treatment of the general planetary motions are his emphasis on observation and his special concern for precise definitions. For example, Ptolemy explained his use of the word "equator" (ἰσημερινός); it is the greatest of the parallel circles described about the poles of "the sphere which rotates all things uniformly" and has this name because it is the only parallel circle which is always bisected by the horizon, and because "the revolution which the Sun makes when located on it produces everywhere what is sensibly" equal day and night.

Ptolemy listed his reasons for supposing the two motions to be as he described; his reasons were based on observation. He explained that the characteristics of the first motion are observed in a single day when whatever celestial objects which are seen rise, cross the meridian (culminate), and set at places which, to the senses, lie on circles parallel to the celestial equator. But the characteristics of the second motion only become apparent after observation over a longer period of time. Only by such long-term observations does it become apparent that the Sun, Moon, and other planets make complicated and differing motions, generally moving toward the east with regard to the stars, which maintain their respective distances and move in a single sphere.

He then explained why it is not sufficient to assign only the first primary motion to all of the celestial bodies. Admitting that it might seem possible that these complicated motions of the planets are due to the motions lagging behind, rather than to an opposite motion, he pointed out that in addition to their movement toward the east, the planets, in their motions, deviate to the north and south. This deviation cannot be explained by suggesting something displacing the motions of the planets, because it is irregular with regard to the primary motion, but is regular with regard to a circle inclined to the equator. Ptolemy went on to explain this second great circle, the λοξός circle (the ecliptic), which describes the second primary motion, and is the same for, and particular to, all the planets. He explained that it is defined by the motion of the Sun, but is also followed by the Moon and the planets. Thus, the second major motion takes place around the poles of a second great circle, inclined to the circle of the primary motion.

Ptolemy then went on to define and describe the equinoctial and tropical points, as well as the meridian circle, in geometrical language. The equinoctial and tropical points are determined by the intersections of the great circles already mentioned, while another great circle, the meridian ($\mu\epsilon\sigma\eta\mu\beta\rho\iota\nu\acute{o}s$), is defined as being orthogonal to the horizon, and determines the midpoint between day and night.[79]

Ptolemy's conception of two primary, but different, motions in the heaven may have been influenced by Plato's conception of the motions of the Same and the Different, described in the *Timaeus* (36b–d). According to Timaeus (38c–e), the Demiurge, when he created the world, split the stuff of the Soul into two strips which he joined in the middle and bent into two rings, naming the outer ring "the Same" and the inner ring "the Different." He assigned to each a particular turning motion, giving supremacy to the revolution of the Same and uniform, which he left single and undivided. The inner revolution was divided in six places into seven unequal circles, and on these were placed the Moon, Sun, and five planets. The *Timaeus* may have been a source for Ptolemy's hypothesis that there are two primary motions in the heavens. This idea may have been attractive to Ptolemy because of its simplicity; it may even have been commonplace.

However, the idea of two primary heavenly motions had been rejected by Aristotle. He explained that "the mathematicians" had described a relationship between the distance of the planets from the first motion and the speed of their motion. He noted that the outermost revolution of the heavens is simple and fastest, while the revolutions of the other are slower and composite. Furthermore, the inner revolutions move in a motion opposite to the circle of the heavens. Nevertheless, Aristotle did not regard these two opposing motions as "primary." Instead, he acknowledged only one "primary" motion, the circular motion of the fixed stars.[80]

Ptolemy's detailed, empirical account of the two primary motions in the heaven, along with his definition of terminology, concludes his summary of the physical concepts preliminary to *The Mathematical Syntaxis*.

✴ 4
Ptolemy's Cosmology

The hypotheses presented in Book One of *The Mathematical Syntaxis* represent Ptolemy's most systematic treatment of physical matters. Nevertheless, these hypotheses do not convey a full picture of Ptolemy's conception of the cosmos. For that, we must look further into the *Syntaxis* and also examine two later works, the *Planetary Hypotheses* and the *Tetrabiblos.*

In the first chapter of Book Nine of the *Syntaxis,* Ptolemy addressed the question of the order of the planets. Ptolemy explained that "almost all the foremost astronomers agree" that all the planets are closer to the Earth than they are to the sphere of the fixed stars, and are farther from the Earth than the Moon. The order of the three outermost planets, Saturn being the farthest, then Jupiter, then Mars, was generally agreed upon. The Moon was the closest celestial body to the Earth. According to Aristotle, who otherwise was not particularly interested in the

106 PTOLEMY'S UNIVERSE

details of planetary order, those planets closest to the outermost sphere move the most slowly;[1] the order for the Moon, Mars, Saturn, and Jupiter was easy to establish this way. But the lengths of the longitudinal periods are the same for Mercury, Venus, and the Sun, being one year, and so are not useful in determining their order.

If the distances of the planets from the Earth were known, their order could be determined. Ptolemy had calculated the distance of the Moon on the basis of observations of lunar parallaxes. However, no planetary parallaxes were observable and so this method could not be used. Ptolemy had determined the distance of the Sun by using a demonstration based upon the relations of the distance of the Sun to the distance of the Moon and the apparent diameters of the Sun, the Moon, and the shadow of the Earth. If a planet passed in front of, or eclipsed, the Sun, that planet must be closer than the Sun to the Earth. Such transits would provide valuable information regarding the order of the planets, if they occurred; however, no such transits had been observed.

Ptolemy related that some earlier astronomers had suggested that the spheres of Mercury and Venus are above that of the Sun; he rejected this suggestion for two reasons. First, even though the Sun has never been seen to be eclipsed by planets, some planets may be below the Sun, but may never be in one of the planes through the Sun and our viewpoint, but in another plane, and therefore would not be seen passing in front of it. Second, Ptolemy approved of using the Sun as a natural dividing line between those planets which may be at any elongation from the Sun and those, namely Mercury and Venus, which always move in its vicinity, but are not close enough to the Earth for parallax to be observed.² The planetary order adopted by Ptolemy was: Moon, Mercury, Venus, Sun, Mars, Jupiter, Saturn.³ He acknowledged that this order was the same as that given by some earlier mathematicians, but he gave no reason for placing Venus above Mercury.

While Ptolemy did not mention any names here, there was by no means unanimity among ancient writers regarding the planetary order. In the *Timaeus,* Plato gave the order of the planets from the Earth as: the Moon, Sun, Venus, Mercury, with the order of the other planets left unstated.⁴ Aristotle was unconcerned with the order of the celestial bodies, suggesting that such matters had been adequately discussed by astronomers, whose writings should be consulted on this matter.⁵ While Vitruvius, Pliny, and Plutarch all subscribed to the same order later adopted by Ptolemy, Cicero could not quite make up his mind.⁶

In a later work, the *Planetary Hypotheses,* Ptolemy returned to the topic of the order of the planets.⁷ In the interim period he had given a great deal of thought to the problem, and now revised some of his earlier opinions. By the time he wrote the *Planetary Hypotheses,* Ptolemy had rejected his earlier reliance on conventional wisdom regarding the order of the planets. He now admitted that the "arrangement of the spheres has been a subject of some doubt" up to this time. As he had earlier held in the *Syntaxis,* Ptolemy reiterated that

> the sphere of the Moon is certainly the closest sphere to the earth; the sphere of

Mercury closer to the earth than the sphere of Venus; the sphere of Venus closer to the earth than the sphere of Mars; the sphere of Mars than the sphere of Jupiter; the sphere of Jupiter than the sphere of Saturn; and the sphere of Saturn than the sphere of the fixed stars.

But while he had found the placement of the Sun between Venus and Mars to be "natural" in the *Syntaxis,* Ptolemy now enumerated three different possibilities for the Sun's position:

> either all five planetary spheres lie above the sphere of the Sun just as they all lie above the sphere of the Moon; or they all lie below the sphere of the Sun; or some lie above, and some below the sphere of the Sun.

Ptolemy was unable to choose between the possibilities, noting that "we cannot decide this matter with certainty."[8]

He went on to discuss the difficulties involved in determining the distance of the planets. Whereas eclipse phenomena had been useful in calculating the distances of the Moon and Sun, no such phenomena could be used for the other planets. As he had in the *Syntaxis,* Ptolemy noted that no occultations of the Sun by a planet had been observed. But Ptolemy presented new explanations for the lack of such occultations. First, concerning the absence of any observed transits of Mercury and Venus, here he suggested that the brightness of the Sun would make any transits of Mercury and Venus invisible: "if a body of such small size (as a planet) were to occult a body of such large size and with so much light (as the sun), it would necessarily be imperceptible, because of the smallness of the occulting body and the state of the parts of the sun's body which remain uncovered."⁹ Second, "such events could only take place at long intervals." The infrequency of transits would also account for the lack of observational evidence. Ptolemy made no mention of the arguments he previously presented in the *Syntaxis,* presumably because he considered these newer arguments to be more valid.

Having concluded that no observations were available to be used to establish the distances of the planets, Ptolemy then explained his mathematical determination of the distances. Assuming a planetary order in which only Mercury and Venus lie below the Sun, Ptolemy used previous computations from the *Syntaxis* together with new calculations. Beginning with the minimum and maximum distances for the Moon and the Sun determined in the *Syntaxis,* as well as the ratios of greatest to least relative distances for each planet (inferred from their geometrical models), Ptolemy relied on two assumptions. First, he assumed that the ratios of relative distances of the planets from the center of the universe correspond to the ratios of their absolute distances. Second, he posited that the planetary spheres are nested one within another, with no empty space between them.

From there, Ptolemy presented his cosmic system, establishing absolute distances for each planet, based on inferences from his initial assumptions. Taking the minimum and maximum distances of the Moon to be 33 and 64 Earth radii, and those of the Sun to be 1160 and 1260 Earth radii, Ptolemy invoked his assumption of nested spheres and found that there was room for Mercury and Venus in between the Moon and the Sun. Relying on his order of the planets and using the ratio of least distance to greatest distance for Mercury as 34:88, Ptolemy stated that "it is clear from the assumption that the least distance of Mercury is equal to the greatest distance of the Moon, that the greatest distance of Mercury is equal to 166 earth radii, if the least distance of Mercury is 64 earth radii." Using 16:104 as the ratio of least distance to greatest distance for Venus, and assuming that the greatest distance of Mercury is equal to the least distance of Venus, Ptolemy calculated the least distance of Venus to be 166 Earth radii and the greatest distance to be 1079 Earth radii.

But, he noted that since "the least distance of the Sun is 1,160 earth radii, as we mentioned, there is a discrepancy between the two distances which we cannot account for," that is, between 1079 and 1160. However, by slightly increasing the distance of the Moon, which correspondingly requires that the distance of the Sun be diminished slightly, this discrepancy is eliminated. The distances of the remaining three planets could then be easily calculated following the same pattern, assuming that their spheres are nested. In earth radii, the least distance of Mars is 1260, while its greatest distance is 8820, which is equal, in turn, to the least distance of Jupiter. Jupiter's greatest distance is 14187 earth radii and is the least distance of Saturn, whose greatest distance is 19865 and is adjacent to the sphere of the fixed stars. The greatest-to-least distance ratios used for each planet were 7:1 for Mars, 23:37 for Jupiter, and 5:7 for Saturn.[10]

It is in this way that Ptolemy derived the planetary distances, probably "as an attempt to demonstrate the order of the planets." After all, Ptolemy's earlier adoption of this same order in the *Syntaxis* was based, by default, merely on convention, since no other means of establishing the order was available. Ptolemy had introduced his discussion of the distances by noting that "[w]e began our inquiry into the arrangement of the spheres with the determination, for each planet, of the ratio of its least to its greatest distance." For Ptolemy, knowledge of the planetary distances would make clear their spatial order.[11]

However, Ptolemy went on to explain that the "argument which forces the above-mentioned order of spheres is not entirely based on the distances, but on the differences in their motions as well." In fact, it was the argument based on motions which Ptolemy claimed was "most compelling." Ptolemy stated that

> the sphere of Mercury is adjacent to the sphere of the Moon, for both the spheres of Mercury and the Moon are eccentric, and the eccenter moves about the center of the universe in the direction of the daily rotation, in contrast to the motion of (the centers of) their epicycles; and it follows that these centers lie at apogee and perigee twice in every revolution.[12]

Because of their similarly complex motions, the Moon and Mercury are near one another. Ptolemy accounted for these complexities by noting that the "spheres nearest to the air move with many kinds of motion and resemble the nature of the element adjacent to them." In other words, the motions of the celestial bodies are affected by the nearby elements and their motions.

Ptolemy's use of physical factors to explain the complexities of celestial motions here is quite striking. He had not employed such explanations in the earlier *Syntaxis*. The *Syntaxis* was primarily an exposition of mathematical astronomy; the models of planetary motion presented there are strictly geometrical. The *Planetary Hypotheses* is in many ways an extension of or companion piece to the earlier *Syntaxis*. The *Planetary Hypotheses*, Ptolemy informed us, contains corrections based on additional observations, in some cases necessitating changes in the planetary models.[13] Book One is devoted to a summary of the planetary models in mathematical terms; the discussion of the order and distance of the planets occurs at the end of this book.

At the beginning of the *Planetary Hypotheses* Ptolemy explained that he wanted to present planetary models in a way that would be useful to instrument makers or builders of equatoria. While such devices could be used for rapid computation, it may also be indicative of Ptolemy's interest in the physical representation of his mathematical models of planetary motion. His discussion of the order and distance of the planets indicates that he was not only interested in the models of motion, but in the physical arrangement of the planets as well. Accordingly, the *Planetary Hypotheses* represents Ptolemy's cosmological conception, based on the assumption that his mathematical models of planetary motions have physical reality, not merely limited to the construction of planetaria.

Indeed, Book Two consists of a discussion of celestial physics, the material substance and cause of motion of the celestial objects.[14] Although he stated that the recounting and correction of his predecessors' views was not his aim, his specific mention of the ideas of both Plato and Aristotle is striking, particularly because Ptolemy rarely mentioned others by name. In fact, Book Two is an extended criticism of Aristotle's celestial physics, the Prime Mover, and the system of rollers and unrollers.

Ptolemy began Book Two by noting that while he had already provided an account of the motions of the celestial bodies, based on observations extending over a long period of time, the task remained to describe the form of those bodies. In offering such a description, Ptolemy explained that he sought to comply with what is appropriate to the nature of the bodies of the spheres as well as with whatever principles necessarily belong to the perpetual, unchanging nature.

Cautioning that his account must be taken only as hypothetical,[15] Ptolemy noted his intention to proceed with a description of individual bodies and their spheres only after a more general discussion of the universal appearances. He pointed out that there are two different points of view, the physical and the mathematical, and emphasized that both must be presented.

According to Ptolemy, the physical point of view leads to the assertion that the aethereal bodies are changeless, allow no interference, and are different from one another. While changelessness is their most important feature, the aethereal bodies are also characterized by their round shape. Their activities are those of things which have parts which are similar throughout. Each sphere is moved by the power of the star which is within that sphere; the beams of the stars are able to freely penetrate the spheres. The motion of each sphere is peculiar to that particular sphere and is caused by the power of an individual star. This motion is similar to animal motion, occurring without constraint or violence. Ptolemy emphasized the animate character of celestial motion, noting that such motion relies on the power of a star to cause the motion of bodies which are similar to it, and drew an analogy to the parts of an animal which move together. He stated that because there is no power stronger than this, no interference with this motion can occur.

Turning to the mathematical point of view, Ptolemy pointed out that there are two different ways to describe the celestial motions. One approach is to use complete spheres, both hollow ones enclosing other spheres as well as solid spheres, called epicycles, on which the stars move. Alternatively, only those portions of the spheres on which the motion is accomplished may be used. If such a portion of a sphere is part of an epicycle, it would have the shape of a tambourine; part of a hollow sphere would be like a belt or a ring or a whorl. (Here Ptolemy referred to Plato by name, mentioning his description of the whorl.[16]) Ptolemy explained that from a mathematical standpoint there is no difference between the two types of description, because the motions which would be assumed in the complete sphere could also be connected in the description which utilizes portions of spheres.

Ptolemy then addressed the merits of the two approaches, noting that the physical point of view might seem to lead to the adoption of complete spheres, because the spherical motions would seem to depend on two points or poles. Here, Ptolemy launched his attack against Aristotle's celestial physics. Ptolemy wove together his arguments, which centered on objections related to physical difficulties involving both the poles and the spheres which transmit motions, as well as the corresponding spheres which inhibit the transmission of motion. He also objected to Aristotle's Prime Mover.

At several points in Book Two, Ptolemy seems to have been directly attacking Aristotle's account of the heavenly motions in the *Metaphysics*.[17] Here, the existence of an Unmoved Mover as the primary principle of celestial motion was discussed. Aristotle explained that movers are substances, and that the celestial movements are more numerous than the bodies moved. To produce the various celestial motions, many spheres are required. Separate spheres are used to transmit motion from the Prime Mover to the inner spheres, because otherwise there would be no connection between the innermost and outermost spheres. Additional spheres were used to produce the differing speeds of the spheres moving the individual celesteal bodies, as well as others which counteracted the motions of those already mentioned.

Ptolemy was adamant in his criticism of Aristotle, arguing that

> we need not attribute to the aethereal bodies things which we suppose necessary for the bodies which exist in our realm, and we need not suppose that something which is appropriate for them, but which hinders bodies which exist in our realm, also hinders the heavenly bodies, which are so very much different.

This passage echoes ideas expressed in the *Mathematical Syntaxis,* where Ptolemy anticipated possible objections to the complexity of his models by asking what could be "more dissimilar than the eternal and unchanging with the ever-changing, or that which can be hindered by anything with that which cannot be hindered even by itself?"[18]

Ptolemy ascribed to Aristotle the use of spheres whose poles were attached to surrounding spheres, although Aristotle did not explicitly discuss such poles. Ptolemy attacked the idea that the pole could be the first cause of the turning motion, stating that there should be no difficulty in accepting the idea that the sphere moves itself and that, furthermore, a motionless object cannot be a source of motion. In chapter five, Ptolemy implicitly argued against Aristotle's Unmoved Mover as well, restating that the source of the motion of the spheres lies within the spheres themselves.

Responding to the possible objection that it is difficult to imagine that the spherical motions of the heavens do not occur around a fixed point, Ptolemy suggested that it is much more difficult to imagine what such a point, the pole, would be like in physical terms. He raised several questions regarding the physical nature of such a pole, asking how the vast surfaces of the spheres could be bound to such a pole, which if it is a point, has no size and basically does not exist. On the other hand, if the pole, a point, is established as a body, other problems will arise in describing its material makeup and its physical nature. That each sphere moves itself seemed only reasonable to Ptolemy; he ridiculed the notion that either a pole or a point could be regarded as the cause of the motion of a sphere, stating instead that the stars are animated and move themselves by choice.

Continuing his attack on Aristotle's explanation, Ptolemy invoked a principle of economy and criticized the idea that an entire sphere should be utilized when a portion of such a sphere would do. Likewise, he found the idea of the "rollers" (*Aufwickeln*) and "unrollers" (ἀνελίττουσαι σφαῖραι),[19] those spheres which contribute to the motions of the celestial bodies or counteract the motions of other spheres, to be senseless and absurd, not only because the universe ends up containing such a large number of spheres, but also because they would occupy so much space, without being necessary to the explanation of celestial motions. Finally, he pointed to the difficulties involved in the notion that the outermost sphere communicates motion to the inner spheres and to the impossibility of determining the source of the motion of the outermost sphere itself.

In chapters seven and eight, Ptolemy enlarged his discussion of the animate nature of the celestial bodies. Drawing an analogy between the celestial bodies and birds, he explained that both are examples of living animals who move in a similar fashion. Choosing a familiar example, he noted that birds do not fly through contact with other birds; this would only hinder their motion. Rather, the origin of their flight resides in a vital force which spreads through the parts of their bodies. Likewise, each planet, also a living and animate being, moves itself, imparting motion to those bodies which are related to it by nature. Ptolemy leaves to the reader the task of drawing the implicit conclusion that, just as in the case of the birds, the contact with the rolling and unrolling spheres would serve as a hindrance to the celestial motions. Ptolemy emphasized the independent nature of the motion of each planet, describing how each planet moves freely, yet in a regular fashion. As an illustration, he chose other analogies to living things, this time to the cooperative, but independent, motions of a circle of dancers or of men in a tournament.

He explained that his theory could be illustrated by the construction of an apparatus to demonstrate the eccentric and epicyclic motions. Once again recalling a passage from the *Mathematical Syntaxis,* he cautioned that what looks simple to us may not reflect the true simplicity of the heavens. In Book Thirteen, Ptolemy had addressed the question of the complicated nature of his devices, explaining that all the parts could pass through and be seen through all the other parts, and that "this ease of transit applies not only to the individual circles, but to the spheres themselves and the axes of revolution." He noted that

> We see that in the models constructed on earth the fitting together of these [elements] to represent the different motions is laborious, and difficult to achieve in such a way that the motions do not hinder each other, while in the heavens no obstruction whatever is caused by such combinations. Rather, we should not judge "simplicity" in heavenly things from what appears to be simple on earth, especially when the same thing is not equally simple for all even here.[20]

Following the general discussion of celestial motions in Book Two of the *Planetary Hypotheses,* Ptolemy began his description of the positions and order of the celestial bodies, starting with the sphere of the fixed stars. He then outlined general terminology and principles. The remaining sections are devoted to motions of the various spheres. Because each planet moves independently, Ptolemy detailed the individual mechanisms for each. In the final section, Ptolemy totaled the numbers of spheres required in his account of the universe, which he stated as forty-one. He was able to further reduce the number of spheres, by assuming that each planet participates in the daily motion by its own choice. He congratulated himself on having produced a simpler system than had any of his predecessors.

In spite of Ptolemy's direct attack on many of Aristotle's ideas, some readers have nevertheless found the *Planetary Hypotheses* to contain indications of a latent Aristotelianism. That the planetary spheres have no wasted space in between them struck Toomer as conforming to Aristotelian thinking. Aristotle had denied the possibility of the existence of a void; the Stoics also had claimed that the universe contains no empty space. While there are important differences between the Aristotelian view and that of the Stoics, Ptolemy did not provide enough information to make clear his own conception of the void and its existence.[21]

To some extent Ptolemy's conception of matter does contain Aristotelian overtones. Goldstein, the discoverer of the previously lost text of Part 2 of Book One, described that section as beginning with "a short Aristotelian introduction." The passage referred to reads as follows (1.2.1):

> These are the models of the planets in their spheres. As we have said, there are anomalies

in the motions of the (planetary) spheres not found in the sphere of the fixed stars, for the latter sphere's motion is very close to that of the universal motion, *whose sphere, of necessity, has a simple nature, unmixed with anything, and containing no contrariety at all.* The planets, *all of which lie below the (prime) mover,* move with it from east to west, and also move with another motion from west to east. They move forward and backward, and to the south and to the north, which are the directions of local motion. *Local motion is the first of the remaining motions and things whose nature is eternal have only this kind of motion. The changes and opposition in quality and quantity, and the coming-into-being of things which are not eternal are not like the changes apparent to us in the eternal, for these changes are in the thing itself and its substance.*

(Those sections of the passage which presumably are "Aristotelian" have here been emphasized.) Certainly, this description is in agreement with the general principles which describe celestial objects in *On the Heavens,* while the description of local motion is in agreement with Aristotle's views expounded in the *Physics.* However, the reference to the "prime mover" (note that "prime" is the reading favored by Goldstein) seems peculiar, in light of Ptolemy's later rejection of such a primary cause of celestial motion in Book Two.[22]

Ptolemy's view that the planets move willfully and independently was found by Pedersen to have something in common with that of Aristotle, explaining that as "in Aristotle, the origin of this conception is animistic: the spheres are moved and animated by living forces intimately connected with the spheres themselves." While this description is accurate, a distinction should also be made between the independent motion of Ptolemy's planets and the motion of Aristotle's planets, which is ultimately caused by the Unmoved, or Prime, Mover.[23]

This difference in view regarding the cause of celestial motion has been cited as a crucial difference between Ptolemy's view and that of Aristotle. Neugebauer described Ptolemy's discussion of the causes of motion as being "obviously in opposition to the Aristotelian system." Sambursky noted that the emphasis which Ptolemy placed "on considering each planet as an independent source of motion is essentially directed against the unitary model of Aristotle."[24]

Sambursky went on to suggest that Ptolemy's description of the independence of planetary motions "hints also at a Platonic conception regarding each planet individually as imbued with a soul as the source of its motion." He attributed Ptolemy's view rather loosely to Pythagorean-Platonic influences, stating that there "remained for Ptolemy, in the age of the return to Pythagoras and Plato, the vitalistic hyothesis of a soul as the driving force of each planet." Sambursky also pointed to Galen, a contemporary, as a possible source for Ptolemy's notion of the independent motion of the planets. He described Ptolemy as "a vitalist who attempts to transfer some basic concepts of vitalism to the dynamics of the heavens."[25] However, Aristotle, as well as Plato and Ptolemy, had also regarded the planets as being animated. The issue which distinguishes Ptolemy's view from that of Aristotle is not whether the planets are animated, but, rather, the source from which they derive their motions. For Ptolemy, each planet moved itself; for Aristotle, the Prime Mover was the ultimate source of motion.

There are further differences in Aristotle's and Ptolemy's understanding of the causes of celestial motions. As was noted above, Ptolemy explained the complexities of the motions of the Moon and Mercury in terms of their resemblance to air, the element adjacent to them. Complexities of planetary motion were treated by Aristotle in *On the Heavens,* where he described the Moon, like the Sun and Earth, as having few motions, while Mercury, like the other planets, has many. Aristotle's explanation of the complexities of planetary motion was tentative and full of disclaimers. Based on his view that the celestial bodies are alive, notions of order and equality work to make sure that the burden of celestial motion is shared. There is nothing in common between Ptolemy's view here and that of Aristotle, for they didn't even agree as to which motions were complex and which were simple; much less did they agree on the cause of complexity. The motions of the Moon were regarded by Aristotle as being relatively simple, while Ptolemy regarded them as complex. Ptolemy attributed these complexities to a physical cause, the similarity to the element air.[26]

Ptolemy's explanation of complex celestial motions leads to the questioning of the extent to which he shared Aristotle's vision of the organization and arrangement of the cosmos. According to Aristotle's conception, the Moon and Mercury would be nearest to the element fire, not air, as

Ptolemy claimed, for the natural place of fire for Aristotle is at the extremity of the sublunar region.[27]

Ptolemy's hints that the celestial bodies share something with terrestrial elements and that the sublunar can affect the superlunar seem to contradict the celestial-terrestrial distinction of Aristotle's cosmology. Pedersen found this break with Aristotelian cosmology curious; noting that "Ptolemy obviously disregards the radical distinction between the heavens above and the elementary spheres below."[28] At the very least, Ptolemy was blurring that distinction, if not ignoring it completely.

Aristotle's account of matter and its motion is at the heart of his differentiation between the two regions of the cosmos; each region has its own type of matter and motion. Aristotle did not refer, except in passing, to any relationship or interaction between these two regions. In *On the Heavens* (289a20–35), Aristotle explained that the heat and light of the stars, including the Sun, is the result of their motion creating friction which causes the air below to be ignited. This is the reason we feel hot when the Sun is near or overhead, because the air is especially heated. Here, Aristotle attributes a change in terrestrial realm to the motion of the celestial bodies.

There are also several passages in the *Meteorology* which are relevant. At both 340b12–14 and 341a13–36, heat felt in the terrestrial region is described as being caused by celestial motions, particularly that of the Sun. Similarly, the heavenly motions help determine the terrestrial winds, for the air surrounding the Earth follows the celestial motions. In fact, at the beginning of the

Meteorology, the celestial motions are credited with being the ultimate cause of all motions in the terrestrial realm:

> This world necessarily has a certain continuity with the upper motions; consequently all its power is derived from them. . . . [We] must assign causality in the sense of the originating principle of motion to the power of the eternally moving bodies.[29]

These passages from the *Meteorology* state unequivocally that the celestial motions cause changes in the terrestrial region.

The passages cited from the *Meteorology* present a much fuller account of interactions between the celestial and the terrestrial regions than do the *Physics* and *On the Heavens.* Except for the passage cited above, the topic is not even broached in the latter two works. Since it is in *On the Heavens* that many of the details of Aristotelian theories of matter and motion are presented, it seems surprising that interactions between celestial and terrestrial matter and motions were not treated there. Of course, there is the possibility that Aristotle had nothing more to say about such interaction.

While passages in the *Meteorology* do mention some effects of celestial motions on terrestrial matter, and one passage may even hint at a blurring of the material distinction between the two regions, Ptolemy went much farther in his blurring of that distinction and his description of interaction between the two regions. For example, there is no suggestion in the *Meteorology* that the sublunar region in any way influences the superlunar, which Ptolemy hinted at in his description of the Moon and Mercury having complex motions because they resemble the air.³⁰ With regard to the question of the influences of celestial motions in the terrestrial region, Ptolemy ultimately went much farther than the *Meteorology* in his treatment of this topic.

An old Greek tradition, that of the *parapegmata,* predicted weather phenomena on the basis of the rising and setting of certain bright stars. This tradition went back to Meton and Euctemon in the fifth century B.C., and may even trace its beginning to Hesiod. Theophrastus's work *On Weather Signs* shows that he was at least aware of the tradition. Ptolemy made his own contribution to the *parapegmata* tradition in his *Phases of the Fixed Stars;* this work also provides important historical information regarding his predecessors within this tradition. Because the *parapegmata* tend to be simply lists of star phenomena correlated to weather, any physical explanation of such correlations between celestial and terrestrial events is lacking.³¹

The idea of a correlation between celestial and terrestrial phenomena was explored further by Ptolemy in the work known as the *Harmonics.* While the work is composed of three books, there are actually two distinct parts. The first, which includes Book One and Two and the first two chapters of Book Three, is concerned with *harmonia* strictly within the context of music.

Ptolemy explained that theoretical knowledge of *harmonia* is a type of mathematics and is concerned with the "ratios of differences between things heard." At the beginning of the *Harmonics,* Ptolemy had hinted at the relationship between astronomy and harmonics. He stated that the aim of the student of harmonics must be to preserve the hypotheses of the *kanon* (a stringed instrument), just as the astronomer's aim is to preserve the hypotheses "concerning the movements of the heavenly bodies in concord with their carefully observed courses."[32] In the rest of Book Three, Ptolemy examined *harmonia* in a wider context.

In this section of the *Harmonics,* Ptolemy set as his goal the investigation of those things which are linked to the power (*dynamis*) of *harmonia.* He explained that this power is a cause, imposing appropriate form on underlying matter. Furthermore, the power of *harmonia* is a rational cause, concerned with the Good, and with the proportions of movements. While this power is necessarily present to some extent in all self-moved things, it is present to the greatest extent in those things which have a more complete and rational nature. Among divine things, the movements of the heavenly bodies are most perfect and rational; among mortal things, the human soul.[33] Ptolemy's task in the *Harmonics* was to identify the analogous structures and relationships present in music, in the heavens, and in the human soul; the same formal, mathematical relationships underlie and define the Good in each type of matter.

The first section of the *Harmonics* presents a detailed description of the musical structures and relations; those of the human soul and of the celestial motions are dealt with more briefly in Book Three. Three chapters (chapters five through seven) treat the soul and human character. In addition to outlining the various parts of the soul and the types of human character in terms of their analogues to musical structures and relations, Ptolemy discussed at some length the relationship between music itself and the activities of the soul. He noted that

> our souls are quite plainly affected in sympathy with the actual activities of a melody, recognising the kinship, as it were, of the ratios belonging to a particular kind of constitution, and being moulded by the movements specific to the idiosyncrasies of the melodies, so that they are led sometimes into pleasures and relaxations, sometimes into griefs and contractions.

He elaborated on the various ways in which the human soul may be affected by hearing music, explaining that "the melody itself modulates in different ways at different times, and draws our souls towards the conditions constituted from the likenesses of the ratios." He suggested that knowledge, possibly intuitive, of this effect was quite ancient, explaining that

> it was because he understood this fact that Pythagoras advised people that when they arose at dawn, before setting off on any activity, they should apply themselves to music and to soothing melody, so that the confusion of their souls resulting from arousal out of sleep should first be transformed into a pure and settled condition and an orderly gentleness, and so make their souls well-attuned and concordant for the actions of the day.

He enlarged on the use of music in a religious context, adding that it "also seems to me that the fact that the gods are invoked with music and melody of some sort . . . shows that we desire them to listen to our prayers with kindly gentleness."[34]

The remaining chapters of the work are devoted to the task of showing that the hypotheses of the celestial bodies are determined in accordance with harmonic ratios. As he did in the *Planetary Hypotheses,* he chose first to treat the general, more comprehensive topics, followed by the more specific. He noted that the musical notes and the heavenly bodies experience only local motion. Furthermore, the motion of the heavenly bodies is circular, while the notes in the musical scales may also be regarded as recurring cyclically. He drew an analogy between a twelve tone scale and the astrological aspects, stating that the fact that

> nature gave the circle of the signs of the zodiac a twelve-part constitution, is what we should rationally expect, since the complete *systēma* [scale] of a double octave is very close to twelve tones, and the interval of a tone was fitted to the twelfth part of the circle.

Turning from circular motion, he then considered three kinds of difference between celestial motions, correlating each to a musical counterpart. He described longitudinal variations (κατὰ μῆκος) as being analogous to variations of pitch, vertical variations (κατὰ βάθος) in "depth," *i.e.,* distance, as correlated to the musical *genera* (the enharmonic, the chromatic, and the diatonic), and variations in latitude (κατὰ πλάτος, here being deviation from the equatorial plane, *i.e.* declination) as being associated with the *tonoi,* or modes. Finally, planetary and lunar phases are correlated with tetrachords and tones in the perfect scale. The last three chapters of the work are no longer extant, although the chapter titles have survived. Strikingly, Ptolemy said nothing here about any interactions between the celestial motions and their terrestrial counterparts in music and the human soul.[35]

It was in the work familiarly known as the *Tetrabiblos* that Ptolemy dealt most extensively with the relationship between and influence of the celestial on the terrestrial. In the opening lines of the *Tetrabiblos,* Ptolemy explained that there are two types of astronomical prediction. One precedes the other and is more effective. This primary type, which allows predictions of the configurations of the Sun, Moon, and stars, was the topic of the earlier *Syntaxis.* The second type of astronomical prediction, which deals with the nature of those configurations and the investigation of the changes which result, is the topic of the *Tetrabiblos.*

This branch of astronomy, referred to as astrology, appears to have been introduced into the Greco-Roman world during the Hellenistic period, perhaps during the second century B.C., becoming particularly important during the Roman imperial period. Hellenistic astrology was particularly concerned with the prediction of human events. There is no evidence that Aristotle knew anything about astrology, and only slim evidence for knowledge on the part of Theophrastus.[36]

Many Stoic philosophers, with the notable exception of Panaetius, were keen advocates of astrology. The Stoic doctrine of συμπάθεια (sympathy) provides an explanation of how such influence can occur physically. According to this conception, all things in the universe participate in cosmic sympathy. The sympathy of the parts which comprise the whole of the world allow the disposition of particular parts to be known, if the disposition of any part is known, for every event affects every other. In this way, future events can be predicted.[37]

At the beginning of the *Tetrabiblos* Ptolemy

stated that it is "evident that most events of a general nature draw their causes from the enveloping heavens." He elaborated on this:

> A very few considerations would make it apparent to all that a certain power emanating from the eternal ethereal substance is dispersed through and permeates the whole region about the earth, which throughout is subject to change, since, of the primary sublunar elements, fire and air are encompassed and changed by the motions in the ether, and in turn encompass and change all else, earth and water and the plants and animals therein.[38]

The four books of the *Tetrabiblos* are devoted to the elucidation of the changes which occur on Earth which can be predicted by the motions of the heavenly bodies.

Book One of the *Tetrabiblos* begins with a general discussion asserting that astronomical prognostication is possible and enumerating its benefits. The bulk of the book is devoted to the presentation of the fundamental concepts, as well as the terminology, of astronomical prognostication. The second book concentrates on general events, such things as climate and natural catastrophes, which relate to the population at large in a given region. Books Three and Four are concerned with the events which affect individuals. Book Three is devoted to the time of birth (nativity), which determines the specific traits of each person. The final book deals with external events which occur in an individual's life, such as the acquisition of fortune, marriage, children, and death.

In all cases, terrestrial events can be predicted

by the physical influence of the motions of the celestial bodies. In Book One, Ptolemy gave a sketch of the primary physical influences of each of the planets. The Sun's effect is to heat and, to some extent, to dry, while the Moon humidifies. Saturn cools and also, to some extent, has a drying effect. Mars has a drying, as well as a burning, effect. Jupiter heats and humidifies, as does Venus, though Jupiter's effect is largely one of heating, while Venus's primary nature is to humidify. Mercury sometimes has a humidifying effect, at other times a drying effect, depending on its changing relationship to the Moon and Sun. In fact, the effects of all of the planets are altered to some degree by their changing positions with regard to the Sun and Moon.

The positions of the celestial bodies, particularly the order and distance of the planets, is significant, for the influence of a planet is mediated by its distance from Earth. The closer the Sun approaches to the zenith, the more noticeable is its heating and drying effect. The Moon humidifies because it is close to the Earth, which produces moisture. Yet, the Moon also contributes a heating effect, due to the light received from the Sun. The cooling and drying effect of Saturn is due to its distance both from the heat-producing Sun and the moist Earth. Because it is so close to the Sun, Mars also produces heat and dryness. Situated between the heating effect of Mars and cooling Saturn, Jupiter's influence is temperate. Venus's heating power is due to its proximity to the Sun, yet the Moon's and Earth's humidity contribute to Venus's humidifying power. So, the physical effects of each celestial body are determined by their physical relationship to one other and their distance from the Earth. The notion that the distances of the celestial bodies from Earth mediate their influence on the terrestrial realm may have been Ptolemy's original contribution.[39]

Because of the importance of the order and distance of the planets in the *Tetrabiblos*, knowledge of the celestial motions investigated in the *Syntaxis* is not sufficient for the type of astronomical prognostication which the *Tetrabiblos* is meant to address. If the physical effects of the planets are to be understood and predicted, their spatial order must also be known. Of course, Ptolemy had stated the order of the planets in the *Syntaxis*. But, within the context of the *Syntaxis*, the order was of no consequence. However, within the *Tetrabiblos* the order is crucial, if the physical effects of each body are to be properly accounted for and predicted. Toomer was rather disdainful of Ptolemy's discussion of the physical influence of the celestial bodies. He explained his point of view, noting that

> Ptolemy regards the influence of heavenly bodies as purely physical. From the obvious terrestrial physical effects of the sun and moon, he infers that all heavenly bodies must produce physical effects (that such an argument could be seriously advanced reflects the poverty of ancient physical science).[40]

Nevertheless, Ptolemy's serious regard for the task of adequately explaining the physical effects of the planets should not be denied. With the writing of the *Tetrabiblos* and its fundamental reliance on an established ordering of the planets, Ptolemy had an increased motivation to justify the planetary order which he had previously adopted rather casually in the *Syntaxis*. The detailed demonstration of the planetary order in the *Planetary Hypotheses* served to fortify the foundation of the physical claims in the *Tetrabiblos*.

These three works are therefore linked in an important way. While no one work of Ptolemy presents his complete views on cosmology, the unity of the several works considered here does provide a partial understanding of his conception of the cosmos. Ptolemy's cosmological ideas can be summarized very briefly, remembering that he himself did not labor over their elucidation.

The Earth is approximately spherical and is very small with regard to the size of the universe as a whole. It is in the middle of the universe, kept there by pressure, probably the result of earthy objects pressing towards the center of the universe. The Earth does not move locally at all.

Terrestrial matter is corruptible, made from imprecisely rounded (but not perfectly spherical) shapes, which are anhomoeomerous. Terrestrial motion was barely mentioned by Ptolemy. He did note that both the direction and paths of proper motion of those bodies having weight is always and everywhere at right angles to the rigid plane drawn tangent to the impact.

The heavens and the heavenly objects move spherically, having two primary motions (the daily motions from east to west, and the motions of the planets along the ecliptic, from west to east). The heavens are greater than any other body, yet the aether has smaller and more rarefied parts than other matter; these parts are spherical, divine, geometrically precise, and physically homoeomerous. (Ptolemy did not explain what he meant by "parts," especially with regard to the divisibility of matter.)

There is no directionality in the universe, no "up" and "down" with respect to the cosmos as a whole. The order of the planets, working outward from the Earth, is Moon, Mercury, Venus, Sun, Mars, Jupiter, and Saturn. There is no empty space between the planetary spheres. Each celestial body is animate, self-moving, and moves independently of other bodies, having its own idiosyncratic motion. The motions of celestial bodies are affected by the adjacent terrestrial elements. The celestial motions influence terrestrial motions and events. Formal, mathematical relations underlie the celestial motions, allowing mathematicians a special opportunity to gain knowledge. Such mathematical relations also underlie music and the human soul.

In considering how little detailed explanation Ptolemy provided to elucidate his cosmological ideas, it must be remembered that physics (what Ptolemy termed "guesswork") was not his real task. Rather, as stated in the preface to the *Syntaxis,* he was primarily interested in the study of mathematics, particularly astronomy, a study which he regarded as an ethical endeavor.

✳ 5
The Divinity of the Celestial Bodies and the Ethical Motivation for the Study of the Heavens

At the beginning of the *Syntaxis,* Ptolemy explained why it was that he had chosen to study astronomy. He was particularly attracted to the study of "the divine and heavenly" (τὰ θεῖα καὶ οὐράνια) and he believed that astronomy "alone is devoted to the investigation of the eternally unchanging." Mathematics, of which astronomy is a type, is the best science to help theology along its way, since it is the only one which can make a good guess at that activity which is unmoved and separated; it is familiar with the attributes of those beings which are on the one hand perceptible, moving and being moved, but on the other hand eternal and unchanging.[1]

Therefore, by concentrating on astronomy, Ptolemy was dedicating himself to the study of the eternally unchanging, divine, and heavenly bodies.

For Ptolemy, the celestial bodies were divine. Furthermore, for him, the study of the celestial motions was an ethical endeavor; the task of the astronomer was great and divine.[2] He was confident that the study of the heavens could result in positive ethical changes for man, noting that "with regard to virtuous conduct in practical actions and character, this science, above all things, could make men see clearly." By studying the celestial motions and emulating them, insofar as possible, man would become more like the divinities. Ptolemy stated that from "the constancy, order, symmetry and calm which are associated with the divine, it [their study] makes its followers lovers of this divine beauty, accustoming them and reforming their natures, as it were, to a similar spiritual state." He concluded the preface to the *Syntaxis* by explaining that it is "this love of the contemplation of the eternal and unchanging which we constantly strive to increase."[3] The conviction that the study and teaching of astronomy would have ethical consequences clearly motivated Ptolemy in his work.

This same ethical theme recurs in his later work as well. At the beginning of the *Tetrabiblos*, Ptolemy was at pains to argue in favor of the value of astrological prediction. He countered the hypothetical argument that foreknowledge of events is superfluous by stating that

> we should consider that even with events that will necessarily take place their unexpectedness is very apt to cause excessive panic and delirious joy, while foreknowledge accustoms and calms the soul by experience of distant events as though they were present, and prepares it to greet with calm and steadiness whatever comes.[4]

Ptolemy regarded astronomy, be it the type found in the *Syntaxis* or that of the *Tetrabiblos,* as providing man with a way to order his life and achieve peace and tranquility.

In his search for such inner peace, Ptolemy shared the ethical goals of many Hellenistic philosophers. While his emphasis on astronomy as a key to ethics was unusual, this point of view was not entirely unique in the history of Greek philosophy, having been advocated by Plato many centuries earlier. Regarding the reverence of the celestial bodies as divinities, Aristotle had been a strong advocate of this view as well. Ptolemy's eclectic views represent an amalgam; his conviction that the study of the divine celestial bodies would produce beneficial ethical effects was the result of various influences found within the Greek philosophical tradition.

The celestial bodies were also regarded as divinities within several near-eastern religious traditions. While there is evidence that these religions influenced Greek ideas regarding celestial divinity, the nature and extent of this influence is not well known, neither in general terms nor with regard to the possible influence of local religious practice on Ptolemy himself. Even though the general subject will not be treated here, several important differences should be pointed out between the Hellenistic astral religions and Ptolemy's point of view. Neither the emphasis on ritualized practices nor the preoccupation with salvation and the after-life which characterize the astral religions are present in Ptolemy's writings. Rather, Ptolemy emphasized his ethical motivation to study the divine celestial motions, stressing that this intellectual endeavor itself is a striving for a noble and disciplined character, which enables the astronomer to become more similar to the celestial divinities.[5]

I. The Greek Cosmological Tradition and the Search for the Divine

When the word *cosmology* connotes the study of the structure of the universe, *cosmos* is taken to refer to the universe as a whole. However, in a more general sense κόσμος (cosmos) means "order" or "adornment" (related to the English word "cosmetic"). The noun is related to the Greek verb κοσμέω, which Gregory Vlastos has so vividly described as

> what the military commander does when he arrays men and horses for battle; what a civic official does in preserving the lawful order of a state; what a cook does in putting foodstuffs together to make an appetizing meal; what Odysseus' servants have to do to clean up the

gruesome mess in the palace after the massacre of the suitors.⁶

The noun κόσμος, related to the verb κοσμέω, gradually developed a restricted sense, being applied to the universe as a whole. Guthrie explained this development, suggesting that

> to a Greek thinker, the most notable thing about the universe was the order which it displayed (above all in events on a cosmic scale like the movements of sun, moon and stars), and this was what contrasted it most radically with the chaos which he supposed to have preceded it, [therefore] the word took on in addition the special meaning of 'world-order' and then simply 'world'.

This change in meaning was gradual and is not easily traced. There are some passages in which it is difficult to determine the sense in which the word is used; it is likely that κόσμος only came to regularly mean "the universe" in the fifth century B.C.⁷

The connection between this later sense of "universe" and the earlier meaning of "order" or "ornament" is highly significant for, even in the restricted usage of "the universe," the idea of order and beauty persisted in Greek thinking. Furthermore, the sense of beauty was not merely a physical aesthetic, but a moral judgment as well. As Vlastos explained, "for the Greeks the moral sense merges with the aesthetic: they commonly say *kalos,* "beautiful," or *aischros,* "ugly," to mean *morally admirable* or *repugnant.*"⁸ Thus the world known as κόσμος has aesthetic, moral, and physical characteristics, all a part of a single idea.

There was a trend in early Greek thought, both before and after the emergence of the idea of the cosmos, to search for, label, and define something in the universe which is divine ($\theta\epsilon\hat{\iota}ον$). For the Greeks, the primary characteristic of their divinities was that they were immortal ($\dot{α}θάνατος$). The earliest portraits of the traditional Greek gods are found in the Homeric and Hesiodic poems. The Homeric poems, with their formulaic language, often refer to the gods as immortal and ageless ($\dot{α}θάνατος\ καὶ\ \dot{α}γήρως$).[9] The gods play important roles in the Homeric poems, only one of which is manifested in their close association with the various phenomena of the sky. Such phenomena are often either named as gods or are depicted as being controlled by gods. The gods hold up the heavens; Zeus can cause night, stars, and various types of weather to appear; Dawn is depicted as a goddess; the Sun is a god. The gods of the Hesiodic *Theogony* parallel and complement the descriptions of the gods given in the Homeric poems.

These early descriptions are the first indication that celestial phenomena will have divine associations in Greek thought, yet further evidence of the divinity of the celestial bodies is lacking until the fifth century. For the most part, the earliest Greek philosophers concerned themselves with new and innovative proposals of divinity.

By the fifth century certain new characteristics of divinity have been added to the catalog. In addition to immortality, divinities may now also be described as the product of some ordering principle, in motion, and spherical. There is a return to the notion of the divinity of the celestial bodies described in the earliest Greek texts.

Philolaus, the fifth-century Pythagorean, is reported to have described ten celestial bodies as divine: the sphere of fixed stars, the five wanderers, the Sun, the Moon, the Earth, the counter-earth, and the central fire. Other evidence of fifth-century belief in the divinity of the celestial bodies includes the allegation that Anaxagoras was put on trial for impious beliefs concerning the celestial bodies. According to Diogenes Laertius, there were many accounts of Anaxagoras's trial for impiety; one of these reports that he was prosecuted for impiety, because he claimed that the Sun was a red-hot mass of metal.[10]

Just as the Pythagoreans had earlier introduced an ethical system based on their conception of divinity, in the fourth century Plato advocated religious piety and an ethics based in part on the divinity of the celestial bodies. In the *Timaeus* (39e–41a), the heavenly bodies are gods who, together with the traditional gods of mythology, created the other living things of the universe, including man. Plato introduced innovation with regard to ethics, while incorporating certain traditional beliefs, including the mythological gods.

The inherent value of traditional beliefs was strongly advocated by Aristotle, who used such beliefs to validate his own views regarding the divinity of the heavenly bodies. His physics was predicated on the distinction between what in nature is divine, that is, eternal, and what is perishable. The heavens were composed of a single eternal element, the divine aether. Aristotle was at pains to establish that his claim about the divinity of the aether was supported by traditional beliefs and everyday language. He used a folk etymology to support his views, explaining that

> the name of this first body has been passed down to the present time by the ancients, who thought of it in the same way as we do, for we cannot help believing that the same ideas recur to men not once nor twice but over and over again. Thus they, believing that the primary body was something different from earth and fire and air and water, gave the name *aither* to the uppermost region, choosing its title from the fact that it "runs always" (ἀεὶ θεῖν) and eternally.

Here, both the motion and the eternity of the aether are depicted as the characteristics which differentiate it from other matter. The weight of traditional thinking is invoked as evidence for the correctness of his view.[11]

Having dealt with the divinity of the aether, Aristotle offered another folk etymology to explain the divinity of the *ouranos* (279a22–279b1), explaining that "the sum of existence of the whole heaven, the sum which includes all time even to infinity, is *aeon,* taking the name from ἀεὶ εἶναι ('to always be'), for it is immortal and divine." Aristotle (284a2–7) placed a high value on the traditional beliefs, explaining that

> we may well feel assured that those ancient beliefs are true, which belong especially to our own native tradition, and according to which there exists something immortal and divine, in the class of things in motion, but whose motion is such that there is no limit to it.

He noted (284a12–14) that "our forefathers assigned heaven, the upper region, to the gods, in the belief that it alone was imperishable; and our present discussion confirms that it is indestructible and ungenerated." Thus, with the writing of Aristotle we have clear evidence of the importance of the belief in the divinity of the heavens.[12]

While it is generally agreed that *On the Cosmos* was not written by Aristotle himself, the work does have the Peripatetic stamp. It also contains some ideas which are not found in Aristotle's writings. For example, *ouranos* is defined as the home of the gods, being "full of divine bodies which we call stars." The eternal motion of the heaven is described (391b14–19) as a "solemn choral dance with all the stars in the same circular orbit revolving unceasingly for all time." The motions of the divine bodies are described and the musical analogy is drawn out:

> The single harmony that is produced by all these as they sing and dance in concert round the heavens has one and the same beginning and one and the same end, in a true sense giving to the whole the name of "order" (κόσμος) and not "disorder" (ἀκοσμία). Just as in a chorus at the direction of the leader all the chorus of men, sometimes of women too, join in singing together, creating a single pleasing harmony with their varied mixture of high and low notes, so also in the case of the god who controls the universe: the note is sounded from on high by him who might well be called the chorus-master; then the stars and the whole heavens move continually.

This passage from *On the Cosmos* recalls a more elaborate account of the dance of the celestial bodies found in the *Epinomis,* probably written by a member of the Academy, if not by Plato himself:

> For mankind it should have been proof that the stars and their whole procession have intelligence, that they act with unbroken uniformity, because their action carries out a plan resolved on from untold ages; they do not change their purpose confusedly, acting now thus, and again thus, and wandering from one orbit to another. Yet most of us have imagined the very opposite; because they act with uniformity and regularity, we fancy them to have no souls. Hence the mass has followed the leading of fools; it imagines that man is intelligent and alive because he is so mutable, but deity, because it keeps to the same orbits, is unintelligent. Yet man might have chosen the fairer, better, more welcome interpretation; he might have understood that that which eternally does the same acts, in a uniform way and for the same reasons, is for that very reason to be deemed intelligent, and that this is the case with the stars. They are the fairest of all sights to the eye, and as they move through the figures of the fairest and most glorious of dances they accomplish their duty to all living creatures.

This imagery was adopted by Ptolemy as well, in the *Planetary Hypotheses* (2.12):

> The parts of the planetary orbits are free to undergo translations and rotations in their natural positions in various ways, except that their movement is uniform revolution, like the chain of hands joined in a circle in a dance, or like the circle of men in a tournament who assist each other and join forces without colliding so as not to be a mutual hindrance.[13]

The image of the cosmic dance may have its source in the ideas of Philolaus, reported by Aetius:

> Philolaus places fire around the centre of the universe, and calls it the 'hearth of the world', the 'house of Zeus', 'mother of the gods', 'altar, bond and measure of nature'. Then again there is another fire enveloping the universe at the circumference. But he says that the centre is by nature primary, and around the centre ten divine bodies dance— first the sphere of the fixed stars, then the five planets, next the sun, then the moon, then the earth, then the counter-earth, and finally the fire of the 'hearth', which has its station around the centre.[14]

The description of the cosmic dance and celestial harmony may have been intended by the author of *On the Cosmos* to serve as a rebuttal to Aristotle's denial of the Pythagorean theory regarding the harmony of the spheres (*On the Heavens* 290b12–291a27).

Nevertheless, Aristotle's description probably served as a model for the description of the aether found in *On the Cosmos:* "The substance of the heaven and the stars we call *aether,* not, as some think, because it is fiery in nature and so burns . . . but because it always moves in its circular orbit; it is an element different from the four elements, pure and divine." The divine aether is described as unchangeable, unalterable, and impassive.[15]

Impassivity was a characteristic shared by the gods of Epicurus and Lucretius. The Epicurean gods had a peaceful, serene existence, unconcerned with human affairs. While Epicurus accepted the existence of gods, and explained man's thoughts about them as *simulacra*, he placed their home in the interstices between worlds. The Stoics did not locate divinity in any single, specific place in the universe. For them, all of nature was divine; there was the idea that the ordering principle of nature was itself divine.[16]

That there is an order which informs the universe, creating a *cosmos*, that wisdom entails knowledge of this order, and that such knowledge is ethical are ideas which shaped Ptolemy's own motivation for practicing astronomy.

II. The Ethical Motivation to Study Astronomy

It is in the *Works and Days* that the first suggestion appears that knowledge of the motions of the celestial bodies may be useful for man, not only for practical purposes, but to help him overcome the evils of the world as well. In fact, the primary purpose of the work as a whole appears to be moral. While "at first sight such a work seems to be a miscellany of myths, technical advice, moral precepts, and folklore maxims without any unifying principle," the poem is composed of four parts which together form a unified whole.[17] The first part of the poem describes the origin and subsequent spread of evil in the world. The second part explains how man may escape these evils through industry, especially in agriculture and trade. A series of maxims useful in everyday life comprises the third section. The final section, which may originally have been a separate work not even written by Hesiod, states which days of the month are favorable for industry and agriculture. The four parts of the poem are linked by their moral aim: to show men how to live in a difficult world. Man may overcome the evil present in this world through industry; astronomical lore allows one to know which days will be particularly favorable for this task.

Many centuries later, this theme was more fully developed in the writings of Plato. Experiencing the disillusionment which many at the turn of the fifth century seemed to feel for the traditional gods and their accompanying morality,[18] Plato sought to establish a religion with visible gods, demonstrating teleological order and an ethics based on the emulation of that order. In the *Timaeus* (39e–41d), the heavenly bodies are gods who, together with the traditional gods of mythology, made the other living things of the universe, including man. However, each human soul had a portion of the immortal soul created by the Demiurge.

It is by means of the celestial gods that ordinary men have several opportunities to gain understanding of the universe. The first opportunity illustrates and depends on Plato's theory of recollection. Before human souls were given their bodies, each soul was assigned to a star, for a ride through the universe, during which the soul was shown the laws of destiny (41d–e). This knowledge, acquired by man before his birth, would later have to be recollected. Once the human soul has acquired its body and begun earthly life, it has other means, also dependent on the celestial bodies by which to acquire knowledge.[19]

Man has been given the power to perceive the

cosmic order for an ethical purpose (*Timaeus* 47b–c):

> God invented and gave us sight in order that we, seeing the revolutions of reason in the heaven, might profit by them for the revolutions of our own intelligence, which are kindred to those, . . . and that learning and sharing in naturally correct reasoning, we might imitate the completely unwandering revolutions of god and stabilize our own wandering motions.

This visibility of the celestial bodies, generated gods, was part of the cosmic plan; the Demiurge made the celestial bodies mostly of fire, so that they might be most bright and fair to see (40a). Unlike the traditional gods of myth who reveal themselves only in so far as they will, the celestial gods are readily seen by all men (40c–41a). This visibility was purposefully ordained by the creator, who, when he ordered the movements of the visible universe, provided a means by which this order could be recognized.

Of the various human sensations, only sight and sound touch the soul. According to Plato, vision is the sensation of motion; man's faculty of vision allows the divinely ordained motions of the heavens to be carried to our souls (47b–c). The sense of hearing also provides access to the divine order (47c–47e; 80b), enabling man to perceive the cosmic harmony, the motions of which are akin to the revolutions of our own souls. Indeed, harmony itself "was given by the gods . . . as an ally for the inner revolution of the soul which has become discordant, to bring it into order and consonance with itself" (47c–d). Once the ordered motions of the universe have been perceived, man is in a position to order his own soul similarly. (Like Plato, Ptolemy also assigned a special status to the senses of sight and hearing).[20]

But while every human has several means by which to imitate the eternal cosmic order, Timaeus claims that only those few who are philosophers will be able to possess the fullest measure of immortality which is appropriate to man (90b–c). The ultimate goal of philosophy itself is ethical, to acquire knowledge of the Good. The philosopher strives for knowledge of true reality (*Republic* 490a–b), which cannot be seen and can only be apprehended by thought (*Republic* 511c–d).[21]

According to the program of education for philosophers outlined in Book Seven of the *Republic,* mastery of dialectic and knowledge of reality can only be acquired after preliminary training in music and gymnastics, followed by a course of mathematical studies which includes arithmetic, geometry, stereometry, astronomy, and harmonics. The type of astronomy to be studied by future guardians is of a very special type and this study will only be undertaken by a few.

In the *Laws* (820e–822d), Plato proposed an educational scheme for the general populace. Although not in training as philosophers, all citizens must pursue a certain amount of astronomical study, not only for the practical value which it may have. The Athenian suggested that citizens learn as much about astronomy as is necessary. For the Athenian, the necessity for study of the heavens was not practical, but religious. He insisted that the citizens and their young people learn enough about the heavenly gods to prevent blasphemy of them, and to ensure piety in sacrifice and prayer. Even if one is not capable of attaining knowledge of the heavens, one should at least acquire true opinion.

This religious value which is placed on the study of astronomy is echoed in the *Epinomis*.[22] The Athenian states that the greatest of human virtues is piety, which he claims can be learned through studying astronomy, for it gives man an understanding of "the generation of divine things, the most beautiful and divine of sights which god has enabled man to see" (991b).

The study of astronomy plays an important role in Plato's ethical philosophy. Motions of the celestial bodies provide the first step in man's ability to recollect the divine knowledge which he learned before he acquired his body. The appearances of these same motions provide other opportunities for all men to live ordered lives, for any man who has either his sight or hearing intact may perceive the divinely ordered motions of the celestial bodies, and thereby order the motions of his own soul accordingly. Plato's educational program for all citizens requires the study of astronomy, as does the more rigorous scheme proposed for philosopher-rulers. For Plato, it is an ethical necessity for all men to be familiar with the celestial bodies and their motions.

Plato presented a strong argument for the place of astronomy within ethical philosophy. Nevertheless, his views regarding the ethical role of mathematics failed to attract the interest of his students. While many of them, including Xenocrates, Speusippus, and Aristotle, did not neglect the study of either mathematics or ethics, they said nothing about the ethical status of mathematics. There is no evidence that mathematicians themselves took an interest in Plato's suggestion. However, in the second century A.D. several influential writers expressed a commitment to Plato's views, and particularly pointed to the beneficial aspects of mathematics.

In their writings, Theon of Smyrna and the author of the *Didaskalikos* each used the language of the mystery cults in their outlines of Platonic philosophy. In their writings, the various branches of mathematics were described as steps in the purification process, necessary preliminaries to initiation into the greater "mysteries" of philosophy.[23] However, their quasi-religious imagery did not contain the strongly ethical component present in Plato's own writings.

The ethical value of mathematics was discussed briefly by Nicomachus, in his *Introduction to Arithmetic,* a primer of Pythagorean number theory containing brief allusions to Platonic concepts. For Nicomachus, numerical relations illustrate the primacy of the beautiful, definite, and intelligible over their opposites. The rational soul orders the irrational, whereby ethical virtues are derived from the resulting equilibrium. The study of arithmetic provides evidence of the order of the universe, for numbers provide cosmic order, as well as orderliness to human life.[24] This emphasis on orderliness was central to Pythagorean number theory and also to that branch of mathematics known as harmonics. That music imparts an ethical influence, aiding man in the cultivation of virtue, is an idea present in Plato's writings, echoed also in the writings of Ptolemy.[25]

The idea that mathematics in general could impart important ethical benefits to those who pursued its study probably originated in the teachings of the Pythagoreans, if not in the writings of Plato.[26] That such a doctrine would appeal to mathematicians should not be surprising; what is surprising is the lack of evidence suggesting that ancient mathematicians adopted this point of view. While Theon, the author of the *Didaskalikos,* and Nicomachus all extolled the benefits for man of studying mathematics generally, it should be remembered that these men were not original mathematicians, but rather popularizers writing elementary handbooks on mathematics. The emphasis on the ethical benefits of mathematics thrived within what was largely a didactic tradition. Plato had stressed the importance of studying all the branches of mathematics; Ptolemy proclaimed his own place within this educational heritage by pointing to his role as a teacher. Strikingly, Ptolemy, as had Plato, described different ways in which an individual could derive the ethical benefits of astronomy, for example, by studying, by teaching, and by making progress in theories.

Of course, Ptolemy was no mere advocate of the study of mathematics, but one of the most influential contributors in the history of astronomy. It is all the more striking that it is in Ptolemy's astronomical writings, which represent a culmination of the Greek mathematical tradition, that we similarly find the culmination of a rather neglected form of Platonic ethical theory, with its special emphasis on astronomy.

By studying mathematics, Ptolemy believed he was pursuing the highest type of philosophy. In his discussion of physical ideas, Ptolemy demonstrated that he was no Aristotelian, but, rather, a mathematician greatly influenced by Platonism. In his self-conscious attempt to outline and explain the fundamental principles and methodological assumptions of his astronomical work, Ptolemy achieved his goal and became a teacher of philosophy. In Ptolemy's universe, mathematics merged with moral philosophy, allowing the mathematician to function as philosopher, all the while striving to imitate the divine.

Notes

Chapter 1: Ptolemy and the Historians

1. *HAMA,* 2: 834. Olympiodorus, *Commentary on Plato's Phaedo,* Lecture 10 section 4, lines 13–15. *Suidae lexicon,* ed. Adler, *s.v.* Πτολεμαῖος, ὁ Κλαύδιος, 4: 254, 3033. The biographical evidence is discussed by Boll, *Stüdien über Claudius Ptolemäus,* pp. 53–66, Ziegler, Ptolemaios," *RE,* columns 1788–1793, and Kunitzsch, *Der Almagest,* pp. 1–2. See *Syntaxis* 10.1 as an example of evidence taken from observations; see also Toomer, "Ptolemy," and *HAMA,* 2: 834–35.
2. *Syntaxis* 2.13.
3. Dillon, *The Middle Platonists,* p. xv.
4. Boll, p. 51.
5. In addition to Lammert's works listed in the bibliography, see also A. A. Long, "Ptolemy on the Criterion."
6. *HAMA,* 2: 940.
7. For examples of the label "eclectic," see Copleston, *A History of Philosophy: Volume I: Greece and Rome Part II,* Brèhier, *The Hellenistic and Roman Age,* and particularly Zeller, *A History of Eclecticism in Greek*

Philosophy, pp. 1–23, *passim*. On religion, see Cumont, *Oriental Religions in Roman Paganism*, and Ferguson, *The Religions of the Roman Empire*.

8. *OCD*, s.v., "Eclecticism."
9. Diogenes Laertius 1.21.
10. Diogenes Laertius 1.14 and 1.19. Harold Cherniss discussed the evidence concerning the nature of the Academy in *The Riddle of the Early Academy*. Sextus Empiricus *Outlines of Pyrrhonism* 1.33 (220), informs us that the founder of the New, or Third, Academy was Carneades. See also Dillon, *Middle Platonists*; Witt, *Albinus and the History of Middle Platonism*; Schmekel, *Die Philosophie der mittleren Stoa*.
11. Zeller, *The Stoics, Epicureans, and Sceptics*, passim; Burnyeat, *The Skeptical Tradition*, passim.
12. Burnyeat, *The Skeptical Tradition, passim*; Zeller, *Stoics, Epicureans, and Skeptics, passim*, but especially pp. 28, 31, 399; Merlan, "Greek Philosophy," pp. 84–106; Burkert, *Lore and Science of Ancient Pythagoreanism, passim*; Dodds, *OCD*, s.v., "Neopythagoreanism."
13. Zeller, *History of Eclecticism, passim*; Dillon, *Middle Platonists, passim*; Merlan, "Greek Philosophy," *passim*.
14. Frazer, *Ptolemaic Alexandria*, 1: 189–301.
15. Two important works on this subject are by Werner Jaeger, *Aristotle: Fundamentals of the History of his Development*, and G. E. L. Owen, "The Platonism of Aristotle."
16. On the Aristotelian corpus, see Porphyry *Life of Plotinus* 24; Plutarch *Sulla* 26. See also Zeller, *Eclecticism*, pp. 113–15. Grayeff's book, *Aristotle and His School*, is in large part concerned with the problem of the authorship of the corpus and provides a short survey of the scholarly literature on the debate. Zeller discussed the same problem and evaluated several theories in *Aristotle and the Earlier Peripatetics*, 1: 105–60. Concerning ancient lists, see that found in Diogenes Laertius 5.22–27. Moraux, *Les listes anciennes des ouvrages d'Aristote*, discussed other lists as well.
17. Eudorus's commentary is now lost. See *OCD*, s.v. "Eudorus," (unsigned article); Dörrie, "Der Platoniker Eudorus von Alexandreia," and Martini, *s.v.* "Eudoros von Alexandrien (10)." On Aspasius, see *OCD*, *s.v.* "Aspasius," (unsigned article); Gercke, *s.v.* "Aspasios (2)." The com-

mentaries of Alexander of Aphrodisias (third century A.D.) on the *Metaphysics* and the *Physics,* as well as those of Themistius (third century A.D.) on the *Physics* and On the Heavens were written too late for Ptolemy to have used them.

18. On Eratosthenes, see Frazer, 1: 409–10. Solmsen, "Eratosthenes as Platonist and Poet," and Knaack, *s.v.* "Eratosthenes von Kyrene (4)." On Atticus, Zeller, *Eclecticism,* pp. 341–43; Freudenthal, *s.v.* "Attikos (18)," *RE.* On Theon see Zeller, *Eclecticism,* p. 339; Ross, "Theon of Smyrna," in *OCD;* von Fritz, *s.v.* "Theon aus Smyrna (14)," *RE.* Concerning Albinus see *Eisagoge,* ed. Hermann, 6: 147–51 and also Alcinous, *Didascalicos,* ed. Hermann, 6: 152–89. Since the nineteenth century, this second commentary has often been recognized as the work of Albinus; see Dillon, p. 268; Dodds, *s.v.* "Albinus," *OCD;* Freudenthal, *s.v.* "Albinus (4)," *RE;* the evidence against this attribution is reviewed by Whittaker, "Platonic Philosophy in the Early Empire."

19. Lammert, "Zur Erkenntnislehre der späteren Stoa," 171–88; Boll, pp. 133–60, 190–94, 206–17; Pedersen, *Survey of the Almagest,* p. 401. Ptolemy referred to the *Syntaxis* at the beginning of the *Tetrabiblos.*

20. Cicero *On Divination* 1.3.6 and *On the Nature of the Gods* 1.3.6; Plutarch *Cicero* 4; Diogenes Laertius 7.138–149; Cleomedes 1.11, ed. Ziegler, pp. 118–19.

21. Edelstein, passage quoted from his surviving papers, reprinted by Kidd in *Posidonius I,* p. xvi. The textual problems and ensuing controversies were discussed by Kidd in his "Introduction," to *Posidonius I,* and by Edelstein in "The Philosophical System of Posidonius," *AJP* 57 (1936): 286–325. Rist, "Categories and their Uses," in Long, *Problems in Stoicism,* p. 44. Edelstein, *AJP* 57: 305.

Chapter 2: The Philosophical Preface to the *Syntaxis*

1. Aristotle *Nicomachean Ethics* 1103a5–19.

2. Aristotle *Metaphysics* 982a2. Here, Aristotle explained that σοφία is ἐπιστήμη about certain principles and causes. Regarding the division of knowledge, Aristotle *Metaphysics* 1025b19–1026a33.
3. Aristotle *Nicomachean Ethics* 1094a19–1095a13.
4. Boll, p. 71; Aristotle *Metaphysics* 1026a6–22; following, with exceptions, Ross's translation. That the divine, for Ptolemy, is the first cause of the first motion of the universe is reminiscent of Aristotle's Unmoved Mover, described at *Metaphysics* 1071b3–1074a38, and also the description of the ὕπατος θεός in the pseudo-Aristotelian *On the Cosmos* 397b24–26. See also Aristotle *Physics* 267a22–b26.
5. The text at 1026a14 is corrupt, with some manuscripts having χωριστά, while others have ἀχώριστα. Aristotle *Metaphysics* 1061b6–8.
6. Aristotle *Metaphysics* 1026a8–10.
7. Aristotle *Physics* 193b22–36; *Metaphysics* 1061a28–1061b4, 1064a32–33.
8. Boll, p. 71.
9. Aristotle *Metaphysics* 1064b1–1064b6.
10. Aristotle *Physics* 260a27–260b20, 265b17–266a5.
11. The use of the word μέσον instead of κέντρον should not mislead us here; μέσον is a perfectly good geometrical term for center, which Ptolemy himself uses in more obviously geometrical passages, *e.g.* at 2.7.
12. Heath, *Mathematics in Aristotle,* p. 1. This book is devoted to an examination of the uses of mathematics in Aristotle. Heath quoted well over one hundred passages from Aristotle's work which use some type of mathematical presentation.
13. Aristotle *Metaphysics* 1061b18–33.
14. *Syntaxis* 1.1, trans. Toomer, *Ptolemy's Almagest,* p. 36.
15. Plato *Symposium* 205d–209c.
16. Plato *Republic* 540a–c.
17. At various places in the *Syntaxis,* for example, his description of his construction of an observational instrument at 5.12 and his detailed descriptions of the arrangement of tables of observations, such as that in 7.4, and his explanation of how to perform various computations, such as the deviation in latitude, explained at 13.6.
18. Plato *Republic* 540b–c; *Symposium* 212a.
19. Plato *Theatetus* 176b.

20. Aristotle *Nicomachean Ethics* 1177a14–17, b31–33; *Didaskalikos* 28.
21. *Didaskalikos* 30.
22. Trans. Toomer, *Ptolemy's Almagest,* p. 37.
23. Lucretius *On the Nature of Things* 5.8–10.

Chapter 3: The Hypothesis Underlying the *Syntaxis*

1. *Syntaxis* 1.2; "hypothesis" is used at 1.8.
2. *Syntaxis* 1.8.
3. Toomer, *Ptolemy's Almagest,* pp. 23–24.
4. Plato *Republic* 510c–511d; *Meno,* particularly 87b–89e. Klein, *Commentary on Plato's Meno,* p. 120.
5. Aristotle *Posterior Analytics* 72a18–24; *Metaphysics* 1061b18–19.
6. Proclus *Commentary on the First Book of Euclid's Elements* Prologue Part 2, ed. Friedlein, pp. 76–77; Euclid *Elements* Book 1, ed. Heiberg, 1: 1–5; pointed out by Morrow, in his translation of Proclus, p. 62, fn. 62.
7. Proclus *Commentary on the First Book of Euclid's Elements* Prologue Part 2, ed. Friedlein, pp. 76–77.
8. It should be noted that outside of these nine chapters, "hypothesis" has other meanings for Ptolemy, most notably "model."
9. Toomer, *Ptolemy's Almagest,* pp. 23–24. Ptolemy's praise of Hipparchus in 9.2, where he called Hipparchus a "great lover of truth" is grounded in his admiration of Hipparchus for having not only compiled observations and investigated theories, but for having argued against the astronomers of his own day because their hypotheses were not in agreement with the phenomena. Ptolemy stated that anyone who is to convince himself and his future audience of his hypotheses must show their agreement with the phenomena. Ptolemy may have admired Hipparchus because he perceived him as sharing a similar attitude, or he may have adopted his attitude regarding the justification of hypotheses by agreement with the phenomena from his reading of Hipparchus.

10. In his discussions of mathematical hypotheses elsewhere in the *Syntaxis,* in some cases Ptolemy was aware that more than one account could be given that would agree with the phenomena. For example, in the case of the Sun's anomaly (3.4), Ptolemy was not only aware that both the epicyclic and the eccentric hypotheses could be used, he provided computations for both. His preference for the eccentric hypothesis was based on his notion of simplicity, discussed in 3.1.

11. *Syntaxis* 9.2.
12. *Syntaxis* 1.3.
13. Ptolemy did not mention any names here. Toomer, *Ptolemy's Almagest,* p. 39, fn. 23, suggested, on the basis of fragment 21A38 in DK (= Aetius 2.13.14), that this idea may have been advocated by Xenophanes of Kolophon (sixth century B.C.). However, as Kirk pointed out in *The Presocratic Philosophers,* p. 173, there is "a divergence in the doxographical accounts of the constitution of the heavenly bodies."

That the celestial bodies move in infinite straight lines was one of the explanations offered by Epicurus, who, of course, did not believe that it was possible to offer a single explanation of celestial phenomena. See his *Letter to Herodotus* 76–82 and *Letter to Pythocles* 90–99, in *Epicurea,* pp. 27–31 and 38–44, and also Asmis, *Epicurus' Scientific Method.* Theon of Alexandria also ascribed this view to Epicurus in his *Commentary, Studi e testi* 72: 338–39.

Pedersen, *Survey of the Almagest,* p. 36, stated his opinion that Theon's ascription of this idea to Epicurus was wrong; he pointed to Diogenes Laertius 10.92 in which he stated that Epicurus "speaks clearly of the rotation of the heavens." Nevertheless, several possible explanations of the rising and setting of the Sun, Moon, and remaining stars ($\tau\hat{\omega}\nu$ $\lambda o\iota\pi\hat{\omega}\nu$ $\mathring{\alpha}\sigma\tau\rho\hat{\omega}\nu$) are also given in the same passage, and are attributed to Epicurus.

14. Aristotle *On the Heavens* 277a12–33.
15. Aristotle *On the Heavens* 277a12–27.
16. Aristotle *On the Heavens* 277a28–33. The difficulty of this passage should be noted. Guthrie, p. 78 of his Loeb ed., commented on the problems of this passage. He was not the first editor to have problems with this passage. Immanuel Bekker, *Aristoteles Graece,* 1:277, 30, accidentally omitted several words, $\epsilon\iota$ δ' $\mathring{\eta}$ $\tau\alpha\chi\upsilon\tau\acute{\eta}s$. Allan, in the OCT edition, proposed the excision of $\tau\alpha\chi\upsilon\tau\hat{\eta}\tau\iota$ from 31.

17. Aristotle *On the Heavens* 273a21–274a18.
18. Simplicius *Commentary on Aristotle's On the Heavens* 1.8 (*CIAG* 7: 266–67) and 4.4 (*CIAG* 7: 710–11).
19. Toomer, *Ptolemy's Almagest*, p. 39, suggested that here Ptolemy was referring to the phenomena of the Sun and Moon appearing larger when close to the horizon. He noted that Ptolemy's explanation here is incorrect, but adds that Ptolemy later correctly explained it as a purely psychological phenomenon in the *Optics* 3.60, ed. Lejeune, p. 116. Toomer claimed, p. 39, that "instrumental measurement of the apparent diameters had convinced [Ptolemy] that the enlargement is entirely illusory." However, the word Ptolemy used in this passage is ἀστήρ, which as Toomer noted, p. 21, can refer to so-called fixed stars or to the heavenly bodies in general. But it is not at all clear from this passage that Ptolemy was alluding specifically to the Sun and Moon.
20. Ptolemy said nothing further here concerning *horoskopia,* which Toomer translated as "sundials." However, in the *Tetrabiblos* (3.2), Ptolemy distinguished different kinds of horoscopic instruments, including solar instruments and water-clocks.
21. On Zenodorus, see Toomer, "The Mathematican Zenodorus;" Heath, *A History of Greek Mathematics,* 2: 207–13. His work is no longer extant; fragments have been preserved in Theon of Alexandria's *Commentary on the Almagest,* Rome ed., *Studi e testi* 72: 355–79.
22. For λεπτομερέστερος, see Aristotle, *On the Heavens* 303b19–22. The superlative form of the word, λεπτομερέστατος, occurs in *On the Soul* 405a6. The aether is discussed in *On the Heavens* 270b1–26. The consideration of generation begins at 298b6–11.
23. However, the word does not occur in the fragments of Anaxagoras. Guthrie, *History of Greek Philosophy,* 2: 325, noted that the

> word ὁμοιομερής does not occur in the extant fragments of Anaxagoras, and since later authorities use it to describe the key-conception of his doctrine one would expect it to be quoted at least once. It does not occur at all before Aristotle, and the idea expressed by it is carefully explained by Plato in the *Protagoras* (329d–e . . .) without the use of the term itself.

See Ross ed., Aristotle *Metaphysics* 1: 132; R. Mathewson, "Aristotle and Anaxagoras," esp. 78 ff.; Barnes, *Presocratic Philosophers,* pp. 321–22.

Aristotle, *On the Heavens* 302a32–b4 (trans. Guthrie, p. 285, Loeb ed.) explained that

> According to Anaxagoras, . . . the homoeomeries are elements (flesh, bone and other substances of that order), whereas air and fire are a mixture of these and all the other seeds, for each consists of an agglomeration of all the homoeomeries in invisible amounts.

For Aristotle's own theory see *On the Heavens* 302b10–20.

24. Theon of Alexandria, *Commentary on the Almagest,* Rome ed., *Studi e testi* 72: 379, 19–21: "ὁμοιομερὲς δὲ τό τε κυκλικὸν ἐν τοῖς ἐπιπέδοις . . . τὸ δὲ σφαιρικὸν ἐν τοῖς στερεοῖς." Mugler, *Dictionnaire, s.v.* ὁμοιομερής, believed that Zenodorus had borrowed this term from pre-Socratic physical doctrine. Theon is the only source which mentions Zenodorus by name. However, two other sources preserve fragments which are very similar to Theon's version. These are: (1) Pappus *Synagoge* 5.4–19, ed. Hultsch, 1: 308–34; and (2) the "Introduction to the *Almagest,*" of which the section on isoperimetric figures is in Hultsch, 3: 1138–64. Mogenet, "L'introduction à l'Almageste," is also helpful here. All three sources may have depended on the same work, perhaps Pappus's lost commentary on Book One of the *Syntaxis,* a suggestion made by Toomer, "The Mathematician Zenodorus." Ptolemy's statement is found at *Syntaxis* 1.3.

25. Proclus *Commentary,* ed. Friedlein, pp. 201, 105, 112–13. Proclus also discussed Geminus's classification of homoeomerous lines at p. 251.

26. Proclus *Commentary,* ed. Friedlein, pp. 112, 120.

27. Aristotle himself rejected what he took to be Anaxagoras's notion of homoeomerity in *On the Heavens* 302b17, noting that: "if what is a compound is not an element, an element is not any homoeomerous body, but that which cannot be divided into constituents which differ in kind." In other words, Aristotle rejected what he took to be Anaxagoras's conception of elements, namely that they contain "seeds" (σπέρματα) of all other things. For Aristotle's use of homoeomeries see *History of Animals* 487a2. Particular aspects of homoeomerous bodies

are discussed in the following passages: in *History of Animals* 486a6 and 489a27, the seat of touch in homoeomerous bodies; *Parts of Animals* 650b13–655b27, the homoeomerous parts of animals, versus heterogeneous parts, like organs; *Generation of Animals* 722a18, heterogeneous parts composed of homogeneous parts; *Generation of Animals* 734b27, the formation of homoeomerous parts. See *Physics* 187a25 on Anaxagoras and infinity; *Physics* 203a21, on the infinity of elements; *On the Heavens* 304a26 concerning the material elements.

28. Mugler, *Dictionnaire* 2:301, s.v. ὁμοιομερής, suggested that both Zenodorus and Proclus borrowed the term from Anaxagoras's doctrines.

29. Theon of Alexandria, *Commentary,* ed. Rome, *Studi e testi* 72: 379.

30. Aristotle *Physics* 260a20–266a9; *On the Heavens* 268b11–269b17.

31. Aristotle *On the Heavens* 293b33–294a10.

32. Aristotle *On the Heavens* 297a8–297b25.

33. Guthrie, Loeb ed., Aristotle *On the Heavens,* p. 244. Stocks, in his notes to his translation of Aristotle *De caelo,* note 1 to 296b20.

34. Aristotle *On the Heavens* 297b20–21: "ἢ οὖν ἐστι σφαιροειδής, ἢ φύσει γε σφαιροειδής."

35. Aristotle *On the Heavens* 297b24–31.

36. Pedersen, *Survey of the Almagest,* p. 39; *HAMA* 3: 1093–1094.

37. Aristotle *On the Heavens* 298a2–6.

38. *Syntaxis* 2.6. Actually, there are several progressions. The κλίματα numbering one through twenty-five are separated by one-quarter of an equinoctial hour (ὥρα ἰσημρινή), so called because these units of time measurement are the same length as one of the seasonal hours (ὧραι καιρικαί) at equinox. A seasonal hour was one-twelfth of the actual length of daytime or nighttime, while the equinoctial hours, used for astronomical purposes, were of uniform length. The κλίματα numbering twenty-six through twenty-nine are separated by one-half equinoctial hour intervals, and κλίματα thirty through thirty-three by one equinoctial hour. Those κλίματα numbered thirty-four through thirty-nine progress at increasing longer intervals as the north pole is approached.

39. *HAMA*, 2: 725, 727.
40. Strabo *Geography* 2.5.34; 1.1.12; *HAMA*, 2: 727; *Syntaxis* 2: 13.
41. Strabo *Geography* 2.2.1.
42. Strabo *Geography* 1.1.20.
43. Strabo *Geography* 1.1.12.
44. Strabo *Geography* 1.1.12; Pedersen, *Survey of the Almagest*, p. 220.
45. *Syntaxis* 2.1; *HAMA*, 2: 938; Toomer, *Ptolemy's Almagest*, p. 75, n. 3.
46. *HAMA*, 2:938; Toomer, *Ptolemy's Almagest*, p. 75. Concerning knowledge of the data, *HAMA*, 2:667–68. Pliny *Natural History* 2.72 (180) gives the second hour of night for Arbela, moonrise in Sicily. Ptolemy, however, in *Geography* 1.4, reports the fifth hour of night for Arbela, the second for Carthage. Neugebauer, *HAMA*, 2:667, noted that the "usefulness of lunar eclipses for the determination of geographical longitudes on the basis of the difference in local time is a pretty obvious consequence of the sphericity of the earth."
47. *Syntaxis* 1.5.
48. This argument was developed more fully by Theon in his *Commentary on the Almagest*, ed. Rome, *Studi e testi* 72: 412ff.
49. Aristotle *On the Heavens* 276a18–b21. At 296b8–9: διὰ τοῦτο γὰρ καὶ τυγχάνει κειμένη νῦν ἐπὶ τοῦ κέντρου. The sense here, conveyed by τυγχάνει κειμένη, is that coincidentally the Earth is at the center of the universe, as a consequence of the motion of the element earth toward the center. Causally considered, the location of the Earth as a whole is a secondary, or incidental, fact. See also 296b35–297a1.
50. See Conroy, *Epicurean Cosmology and Hellenistic Astronomical Arguments*, for a detailed discussion of the Epicurean viewpoint. Lucretius *On the Nature of Things* 5.534–542; on the lack of a privileged place, 1.1052–1090. Although the text is mutilated in places, the general line of reasoning remains clear.
51. See Hahm, *Origins of Stoic Cosmology*, pp. 120–21, and Furley, "Lucretius and the Stoics," p. 20, on the relationship between the Epicurean and Stoic treatments of the problem. Stobaeus *Eclogae* 1.19.4, ed. Wachsmuth, 1: 166 (= Arius Didymus *Epitome* 23, p. 459 *Doxographi Graeci*).
52. Plutarch *On the Face of the Moon* 6–7 (992–924).

53. Plutarch *On the Face of the Moon* 8 (924).
54. *Syntaxis* 1.6.
55. Aristotle *On the Heavens* 297b31–298a21.
56. Aristotle *On the Heavens* 298a6–8.
57. *Syntaxis* 1.5.
58. *Syntaxis* 5.11.
59. *Syntaxis* 5.11.
60. Euclid *Phenomena,* ed. Heiberg, 8: 10; Aristarchus *On the Sizes and Distances of the Sun and Moon* Hypothesis 2, in Heath, *Aristarchus of Samos,* p. [353]; Geminus *Eisagoge* in *Elementa Astronomiae* 17.16, ed. Manitius, p. 186.
61. *Syntaxis* 1.7.
62. Aristotle *On the Heavens* 296a24–296b27.
63. Aristotle *On the Heavens* 296a27–34.
64. Aristotle *On the Heavens* 296a34–296b7.
65. Aristotle *On the Heavens* 296b7–21, 310b3–5.
66. Curtis Wilson has pointed out to me (in a letter) that, strictly speaking, "the rigid plane drawn tangent to the impact," does not quite do the job here. For example, the impact might be on a mountain side. Ptolemy did not consider such possibilities, but described the fall and impact of heavy objects in idealized, geometrical language.
67. Aristotle *On the Heavens* 296b23–25.
68. Aristotle *On the Heavens* 296b32–297a3.
69. Aristotle *On the Heavens* 297a3–8.
70. Plato *Timaeus* 62c.
71. Plato *Timaeus* 62c–63e. Cornford, *Plato's Cosmology,* p. 262 n. 3, provided the following explanation:

> No part has the property of 'being above (or below) the centre', or has any better right to that description than a point on the opposite side. This is the counterpart of the statement above, that the centre cannot be called 'above' the 'lower' hemisphere or 'below' the 'upper' hemisphere.

72. E.g. Aristotle *Physics* 214b14; 212a21–28.
73. Aristotle *On the Heavens* 308a14–21. The comments on direction terminology are subsidiary to Aristotle's consideration of heaviness and lightness. While Aristotle recommended the use of direction terminology which reflects the relative character of heaviness and lightness, he insisted on the existence of absolute heavi-

ness and absolute lightness in *On the Heavens* 311b14–312a21.

74. Plato *Timaeus* 63a.

Cornford's comment on this passage is helpful, *Plato's Cosmology,* p. 263:

> ... the supposed traveller will be using 'above' and 'below' with reference to every direction in succession. ... Neither word accordingly, stands for any inherent difference between the parts of the central body or of the universe as a whole.

75. Aristotle *On the Heavens* 308a21–30.
76. Plato *Timaeus* 62c–63e.
77. Aristotle *On the Heavens* 296b29–297a2.
78. *Syntaxis* 1.8.
79. *Syntaxis* 1.8.
80. Aristotle *On the Heavens* 291a29–291b10, 292a10–14.

Chapter 4: Ptolemy's Cosmology

1. Aristotle *On the Heavens* 291a33–b7.
2. *Syntaxis* 9.1.
3. *Syntaxis* 9.1.
4. Plato *Timaeus* 38d.
5. Aristotle *On the Heavens* 291a29–32.
6. Vitruvius *On Architecture* 9.1.5; Pliny *Natural History* 2.6; Plutarch *On the Generation of the Soul in Plato's Timaeus* 31; Cicero *On Divination* 2.43. Cicero adopted a different order in *On the Nature of Gods* 2.21.
7. There are references in the *Planetary Hypotheses* to *The Mathematical Syntaxis* at 1.1 and 1.2. Neugebauer suggested that the *Planetary Hypotheses* may have been Ptolemy's final effort, *HAMA,* 2: 901. The *Planetary Hypotheses* survives in its entirety only in Arabic and Hebrew translations; only the first part of Book One survives in Greek.
8. Ptolemy *Planetary Hypotheses* 1.2.2 (Goldstein ed.).

9. Trans. Toomer, *Ptolemy's Almagest,* p. 419, fn. 2.
10. Ptolemy *Planetary Hypotheses* 1.2.3 (Goldstein ed.).
11. Ptolemy *Planetary Hypotheses* 1.2.3 (Goldstein ed.).

Swerdlow, *Ptolemy's Theory of the Distances and Sizes of the Planets,* pp. 103–6. It should be noted here that Ptolemy revised his earlier statement that none of the planets display parallax. He now stated *(Planetary Hypotheses* 1.2.5, Goldstein ed.) that if his examination of planetary distances and sizes in the *Planetary Hypotheses* was correct

> Mercury, Venus, and Mars display some parallax. The parallax of Mars, at perigee, is equal to that of the Sun at apogee. The parallax of Venus at apogee is close to that of the Sun at perigee. The parallax of Mercury at perigee is equal to that of the Moon at apogee, while the parallax of Mercury at apogee is equal to that of Venus at perigee. The ratio of each of them to the lunar and solar parallax is equal to the ratio of the distances that we have mentioned to the distances of the Sun and the Moon.

It is important to note here that Ptolemy did not offer observations of parallax, rather the prediction that Mercury, Venus, and Mars would exhibit parallax based on his calculations of their distances.

12. Ptolemy *Planetary Hypotheses* 1.2.3 (Goldstein ed.).
13. Ptolemy *Planetary Hypotheses* 1.2 (Heiberg ed.).
14. As mentioned earlier, Book Two no longer survives in Greek. My reading is based on the German translation from the Arabic by Buhl and Heegaard.
15. Stephen Toulmin and Ernan McMullin each pointed out to me that this caution may be a veiled reference to the "likely account" described by Timaeus at 29d.
16. Plato gave a lengthy description of the whorls attached to the spindle of Necessity, through which the heavenly motions turn, in the *Republic* 616c–617b.
17. Aristotle *Metaphysics* 1073a23–1074a31.
18. *Planetary Hypotheses* 2.5; *Syntaxis* 13.2, trans. Toomer, p. 600.

19. The German translation of Book Two contains the word for "rollers," while the *Metaphysics* (1074a2-14) uses the phrase for "unrolling spheres."

20. *Syntaxis* 13.2; trans. Toomer, *Ptolemy's Almagest,* p. 601.

21. Toomer, *s.v.* "Ptolemy," *DSB,* p. 197; Aristotle *Physics* 217b20-28; on the Stoics see Diogenes Laertius 7.140.

22. Goldstein, Introduction to his translation and edition of "The Arabic Version of Ptolemy's *Planetary Hypotheses,*" p. 4; his translation of 1.2.1 is quoted. For example, *On the Heavens* 268b11-269a18; *Physics* 260a20-261b26.

23. Pedersen, *Survey of the Almagest,* p. 396; Aristotle *Physics* 267a20-b26.

24. Neugebauer, *HAMA,* 2: 923; Sambursky, *The Physical World of Late Antiquity,* p. 142.

25. Sambursky, *The Physical World of Late Antiquity,* pp. 140-45.

26. Ptolemy *Planetary Hypotheses* 1.2.3 (Goldstein ed.); Aristotle *On the Heavens* 291b28-293a13.

27. Aristotle *On the Heavens* 311b22-27.

28. Pedersen, *Survey of the Almagest,* p. 393.

29. Aristotle *Meteorology* 339a20-33, trans. Webster, OCT ed.

30. *Planetary Hypotheses* 1.2.3 (Goldstein ed.).

31. *HAMA,* 1: 4, 2: 588; Hesiod *Works and Days* 417ff., 587f., 609ff.

32. *Harmonics* 1.2, trans. Barker, *Greek Musical Writings II,* p. 278.

33. *Harmonics* 3.3-3.4.

34. *Harmonics* 3.7, trans. Barker, *Greek Musical Writings II,* p. 379.

35. *Harmonics* 3.8-13; trans. Barker, *Greek Musical Writings II,* p. 383.

36. On Hellenistic astrology see Neugebauer, *Exact Sciences in Antiquity,* pp. 187-89; *HAMA,* 2: 608-10; Cramer, *Astrology in Roman Law;* MacMullen, *Enemies of the Roman Order,* chap. 4. Proclus, a very late (fifth century) source, reported in his *Commentary to Plato's Timaeus* Book 4 (*Timaeus* 40c-d, Diehl ed., vol. 3: 151) that according to Theophrastus the Chaldeans predicted life and death as well as the weather.

37. Cicero *On Divination* 2.52-56; 2.14; *On the*

Nature of the Gods 2.7.19; Sextus Empiricus *Adversus Mathematicos* 5; *Adversus Mathematicos* 9.79ff. (*Against the Physicists* 1, in the Loeb ed.)

38. Ptolemy *Tetrabiblos* 1.1–2, trans. Robbins, pp. 5–7.

39. According to Neugebauer, *HAMA*, 2: 609, fn. 13, astrological theory was not concerned with the effects of the planets based on their geocentric distances.

40. Toomer, *s.v.* "Ptolemy," *DSB*, p. 198.

The *Planetary Hypotheses* was also intended to be useful to builders of planetaria. Such models may have been attractive to those attempting astronomical prognostication.

The remarks of G. E. R. Lloyd in *The Revolutions of Wisdom*, p. 43, seem worth repeating in this context:

> The fact that most prominent ancient astronomers, including Hipparchus and Ptolemy, also engaged in astrology is often taken to be irrelevant to Greek *science* and as evidence only of the failure of the Greeks to be *scientific*. Yet not to be guilty of gross anachronism, we must take as our explananda not just those parts of ancient mathematics and natural philosophy that we approve or consider fruitful, but the whole of the corpus of work of those who engaged in different branches of those complex and manifold traditions. To ignore astrology would be to miss the insights it can offer both about ancient controversies concerning what those traditions comprised and about the ambitions some theorists entertained concerning some areas that they were certainly eager to include.

Chapter 5: The Divinity of the Celestial Bodies and the Ethical Motivation for the Study of the Heavens

1. *Syntaxis* 1.1, trans. Toomer, *Ptolemy's Almagest*, p. 36.

2. *Syntaxis* 9.2; 4.9.
3. *Syntaxis* 1.1, trans. Toomer, *Ptolemy's Almagest*, p. 37.
4. *Tetrabiblos* 1.3, trans. Robbins, Loeb ed., p. 23.
5. On near-eastern religions and their influence, see Bouchè-Leclercq, *L'Astrologie Grecque;* Cumont, *Astrology and Religion; L'Egypte des Astrologues;* and *Oriental Religions.*
6. Vlastos, *Plato's Universe,* p. 3.
7. Guthrie, *History of Greek Philosophy,* 1: 110. Examples of the changes in meaning include: Heraclitus fragment 22B30 (= Clement *Stromateis* 4.105); Empedocles fragment 31B134 (= Ammonius *de interpretatione* 249, 6 Busse), in DK. See also Guthrie, 1: 208–9.
8. Vlastos, *Plato's Universe,* p. 3.
9. E.g. *Odyssey* 5.218; *Iliad* 2.447.
10. On Philolaus, DK 44A16 = testimony of Aetius 2.7.7; on Anaxagoras, DK 59A1 = Diogenes Laertius 2.12–13. On Pythagorean practices, see Burkert, *Lore and Science,* pp. 166–92.
11. Aristotle *On the Heavens* 269a30–269b14. The passage quoted is 270b17–25, trans. Guthrie, Loeb ed.; cf. Aristotle *Meteorology* 339b19–30. Plato had a similar interest in etymology; see, *e.g. Cratylus* 410b-c.
12. Aristotle *On the Heavens* 284a2–14, trans. Guthrie, Loeb ed.
13. Pseudo-Aristotle *On the Cosmos* 399a12–21, trans. Furley, Loeb ed.; Plato *Epinomis* 982c–e, trans. Taylor, in *Plato: The Collected Dialogues;* Ptolemy *Planetary Hypotheses* 2.12, trans. Sambursky, *The Physical World of Late Antiquity,* p. 145.
14. DK 44A16 = Aetius 2.7.7, trans. Raven, in Kirk, Raven, and Schofield, *The Presocratic Philosophers,* p. 343.
15. Pseudo-Aristotle *On the Cosmos* 392a6–9, trans. Furley, Loeb ed.; also, 392a31–35.
16. Lucretius *On the Nature of Things* 2. 1090–1104; on simulacra, *On the Nature of Things* Book 4; Epicurus *Letter to Herodotus* 46–52; on the home of the gods, *On the Nature of Things* 3.18–24, 5.146–73. Concerning the Stoics, Cicero *On the Nature of the Gods* 1.15.36.
17. White, "Introduction," *Hesiod: The Homeric Hymns and Homerica,* Loeb ed., p. xviii.
18. Solmsen, *Plato's Theology,* chaps. 1–3.

19. The theory of recollection is first introduced in the *Phaedo* (73c–76b) and the *Meno* (82b–86b). According to the account of Timaeus, the Creator thought it proper to create the number of human souls to be equal to the number of stars.

Socrates's second speech in the *Phaedrus* (246a–250a) coincides in many ways with Timaeus's description of the soul's celestial journey before birth and subsequent reincarnations. In Book Ten of the *Republic* (614b–621d), the myth of Er recounts how souls after death (that is between reincarnations) encounter a model of the astronomical revolutions and harmonies linked to the laws of destiny and then must choose their next life, to be lived in compliance with what has been seen.

20. I am indebted to Richard Baker for pointing out that the sensations of sight and hearing *alone* touch the soul (μέχρι ψυχῆς); see *Timaeus* 67a7–b7. Timaeus explained that a human being does not perceive objects, but external motions, which, being imparted through sense organs, become internal motions (45b–d). When a low sound (a slow motion) overtakes a high sound (a swifter motion), a harmonious sound is perceived which imitates the divine harmony (80a–b) in which, likewise, swift motions are seemingly overtaken by slower motions (described at 39a–b). These harmonious sounds are most pleasant to man. On sight and hearing, also see *Harmonics* 3.3.

21. In the *Republic* (500d), Socrates stated that "the philosopher, being familiar with the divine order, will himself become orderly and divine as much as is possible for man." He explained in the *Phaedrus* that "Reality exists without shape or color, intangible, visible only to reason, the pilot of the soul; true knowledge is knowledge of it" (247c). In the *Symposium,* Socrates claimed that once having ascended the ladder of love, "beholding beauty with the eye of the mind, man will be enabled to bring forth, not images of beauty, but realities . . . and having brought forth and nourished true virtue he will become the friend of God and be immortal, if mortal man may" (*Symposium* 212a).

22. The *Epinomis* may have been written by Philip of Opus. See Tarán, *Academica,* and Taylor, "Plato and the Authorship of the 'Epinomis.'"

23. Theon, Hiller ed., p. 1, lines 1ff.; p. 14, lines 18ff. Alcinous *Didaskalikos* 28, Hermann ed., p. 182, lines 7ff. The author of the *Didaskalikos* is difficult to identify; see Whittaker, "Platonic Philosophy in the Early Empire."

24. Nicomachus *Introduction to Arithemetic* 1.14.2; 1.23.4.

25. Plato *Republic* 522a, *Timaeus* 47c–e, 80a–b. Ptolemy *Harmonics* 1.2; 3.4ff.

26. See Burkert, *Lore and Science,* who argued that much of "Pythagoreanism" was intellectual propaganda disseminated by Plato and his followers.

Bibliography

I. Writings of Ancient Authors (includes editions and translations)

Albinus. In *Platonis Dialogi secundum thrasylli tetralogias dispositi.* Ed. C. F. Hermann. 6 vols. in 3. Leipzig: Teubner, 1859. Vol. 6.

Alcinous. In *Platonis Dialogi secundum thrasylli tetralogias dispositi.* Ed. C. F. Hermann. 6 vols. in 3. Leipzig: Teubner, 1859. Vol. 6.

Aristarchus. *On the Sizes and Distances of the Sun and Moon.* Trans. Thomas Heath. In *Aristarchus of Samos: The Ancient Copernicus.* New York: Dover, 1981, rpt. of Oxford: Clarendon Press, 1913 ed.

Aristotle. *Aristoteles Graece.* Ed. I. Bekker. 3 vols. Berlin: Georg Reimer, 1831.

———. *Analytica priora et posteriora.* Ed. W. D. Ross. Oxford: Clarendon Press, 1964.

———. *The Complete Works of Aristotle.* The Revised Oxford Translation. Ed. J. Barnes. 2 vols. Princeton: Princeton University Press, 1984. Bollingen Series 71, 2.

———. *De anima.* Ed. W. D. Ross. Oxford: Clarendon Press, 1956.

———. *De animalibus historia*. Ed. L. Dittmeyer. Leipzig: Teubner, 1907.

———. *De caelo*. Trans. J. L. Stocks. Oxford: Clarendon Press, 1922.

———. *De caelo libri quattuor*. Ed. D. J. Allan. Oxford: Clarendon Press, 1955, rpt. of 1936 ed.

———. *De generatione animalium*. Ed. H. J. Dressaart Lulofs. Oxford: Clarendon Press, 1965.

———. *Ethica Nicomachea*. Ed. I. Bywater. Oxford: Clarendon Press, 1949, rpt. of 1894 ed.

———. *Historia animalium*. Ed. and trans. A. L. Peck. 3 vols. Cambridge: Harvard University Press, 1965. Loeb.

———. *Metaphysica*. Ed. W. Jaeger. Oxford: 1980, rpt. of 1957 ed.

———. *Metaphysics*. Trans. W. D. Ross. 2 vols. Oxford: Clarendon Press, 1924.

———. *The Metaphysics Books I–IX*. Trans. H. Tredennick. Cambridge: Harvard University Press, 1935. Loeb.

———. *The Metaphysics Books X–XIV*. Trans. H. Tredennick. Cambridge: Harvard University Press, 1933. Loeb.

———. *Meteorologica*. Ed. and trans. H. D. P. Lee. Cambridge: Harvard University Press, 1978, rpt. of 1952 ed. Loeb.

———. *Meteorologica*. Ed. E. W. Webster. Oxford: 1927.

———. *The Nicomachean Ethics*. Ed. and trans. H. Rackham. Cambridge: Harvard University Press, 1982. Loeb.

———. *On the Heavens*. Ed. and trans. W. K. C. Guthrie. Cambridge: Harvard University Press, 1971. Loeb.

———. *On Sophistical Refutations; On Coming-to-be and Passing Away; On the Cosmos*. Ed. and trans. E. S. Forster and D. J. Furley. Cambridge: Harvard University Press, 1955. Loeb.

———. *Parts of Animals; Movement of Animals; Progression of Animals*. Ed. and trans. A. L. Peck and E. S. Forster. Cambridge: Harvard University Press, 1937. Loeb.

———. *Physica*. Ed. W. D. Ross. Oxford: Clarendon Press, 1950; rpt., with corrections, 1982.

———. *The Physics.* Ed. and trans. P. H. Wicksteed and F. M. Cornford. 2 vols. Cambridge: Harvard University Press, 1934. Loeb.

Cicero. *De natura deorum; Academica.* Trans. H. Rackham. Cambridge: Harvard University Press, 1979, rpt. of 1933 ed. Loeb.

———. *De senectute, De amicitia, De divinatione.* Ed. W. A. Falconer. Cambridge: Harvard University Press, 1979, rpt. of 1923 ed. Loeb.

Cleomedes. *De motu circulari corporum caelestium libri duo.* Ed. H. Ziegler. Leipzig: Teubner, 1891.

Diels, H. *Die Fragmente der Vorsokratiker.* Ed. W. Kranz. 3 vols. Berlin: Weidmannsche Verlagsbuchhandlung, 1934, 5th ed.

Diogenes Laertius. *Vitae philosophorum.* Ed. H. S. Long. 2 vols. Oxford: Clarendon Press, 1964.

Doxographi graeci. Ed. H. Diels. Berlin: Georg Reimer, 1879.

Epicurus. *Epicurea.* Ed. H. Usener. Leipzig: Teubner, 1887.

Euclid. *Opera omnia.* Ed. J. L. Heiberg and H. Menge. 8 vols. Leipzig: Teubner, 1883–1916. (*Elementa* vol. 1–5.)

———. *The Thirteen Books of Euclid's Elements.* Trans. T. L. Heath. 3 vols. New York: Dover, 1956 rpt. of 1926, 2nd ed.

Further Greek Epigrams: Epigrams before A.D. 50 from the Greek Anthology and other Sources not included in "Hellenistic Epigrams" or "The Garland of Philip". Ed. D. L. Page. Cambridge: Cambridge University Press, 1981.

Geminus. *Elementa astronomiae.* Ed. C. Manitius. Leipzig: Teubner, 1898.

Herodotus. *Herodotus.* Ed. and trans. A. D. Godley. 4 vols. Cambridge: Harvard University Press, 1960; rpt. of 1920 ed. Loeb.

Hesiod. *Hesiod: The Homeric Hymns and Homerica.* Ed. and trans. H. G. Evelyn White. New York: G. P. Putnam's Sons, 1982; rpt. of 1914 ed. Loeb.

Homer. *The Iliad.* Ed. and trans. A. T. Murray. 2 vols. Cambridge: Harvard University Press, 1976; rpt. of 1925 ed. Loeb.

———. *The Odyssey.* Ed. and trans. A. T. Murray. 2 vols. New York: G. P. Putnam's Sons, 1919. Loeb.

Lucretius. *De rervm natvra libri sex.* Ed. C. Bailey. Oxford: Clarendon Press, 1947.

Nicomachus. *Introdvctionis arithmeticae libri II.* Ed. R. Hoche. Leipzig: Teubner, 1864.

———. *Introduction to Arithmetic.* Trans. M. L. D'Ooge. New York: Macmillan, 1926.

Olympiodorus. *The Greek Commentaries on Plato's Phaedo.* Vol. 1. *Olympiodorus.* Ed. L. G. Westerink. Amsterdam: North-Holland Publishing, 1976. *Verhandelingen der Koninklijke Nedelandse Academie van Wetenschappen,* Afd. Letterkunde, Nieuwe Reeks, deel 92.

———. *In Aristotelis Meteora Commentaria.* Ed. G. Stüve. *Commentaria in Aristotelem Graeca,* 12, pt. 2. Berlin: Georg Reimer, 1900.

Pappus. *Collectionis quae supersunt.* Ed. F. Hultsch. 3 vols. Berlin: 1876–1878; rpt. Amsterdam: Adolf M. Hakkert, 1965.

Plato. *The Collected Dialogues.* Ed. E. Hamilton and H. Cairns. Princeton: Princeton University Press, 1980 rpt. of 1961 ed. Bollingen Series 71.

———. *Opera.* 6 vols. Ed. J. Burnet. Oxford: Clarendon Press, 1979–82, rpt. of 1901–7 ed.

Pliny. *Natural History.* Ed. and trans. H. Rackham. 10 vols. Cambridge: Harvard University Press, revised and reprinted, 1949. Loeb.

Plutarch. *Plutarch's Lives.* Ed. B. Perrin. 11 vols. Cambridge: Harvard University Press, 1928–1950. Loeb.

———. *Plutarch's Moralia XII.* Ed. and trans. H. Cherniss and W. C. Helmbold. Cambridge: Harvard University Press, 1957. Loeb.

———. *Plutarch's Moralia XIII,* Pt. 1. Ed. and trans. H. Cherniss. Cambridge: Harvard University Press, 1976. Loeb.

———. *Vitae Parallelae.* Ed. K. Ziegler. 5 vols. Leipzig: Teubner, 1968.

Porphyry. *Vita Plotini.* In *Plotin. Ennéades.* Vol. 1. Ed. É. Bréhier. Paris: Société d'Édition "Les Belles Lettres," 1924.

Posidonius. *Posidonius I. The Fragments.* Ed. L. Edelstein

and I. G. Kidd. Cambridge: Cambridge University Press, 1972.

Proclus. *In Platonis Timaeum commentaria.* Ed. E. Diehl. 3 vols. Amsterdam: Adolf M. Hakkert, 1965, rpt. of Leipzig: Teubner, 1903.

———. *In primum Euclidis Elementorum librum commentarii.* Ed. G. Friedlein. Leipzig: Teubner, 1873.

———. *Proclus. A Commentary on the First Book of Euclid's Elements.* Trans. Glenn R. Morrow. Princeton: Princeton University Press, 1970.

Ptolemy. *The Almagest by Ptolemy.* Trans. R. C. Taliaferro. Great Books of the Western World 16. Chicago: Encyclopedia Britannica, 1952.

———. "The Arabic Version of Ptolemy's *Planetary Hypotheses.*" Ed. B. R. Goldstein. *Transactions of the American Philosophical Society,* n.s. 57, pt. 1 (1967): 3-55.

———. *Composition mathématique de Claude Ptolémée.* 2 vols. Ed. M. l'abbe [Nicolas B.] Halma. Paris: Henri Grand, 1813-1816.

———. *Geographia.* Ed. C. F. A. Nobbe. Hildesheim: Georg Olms, 1966; rpt. of Leipzig 1843-45 ed.

———. *Handbuch der Astronomie.* Trans. K. Manitius, new edition corrected by O. Neugebauer. 2 vols. Leipzig: Teubner, 1963.

———. *Harmonics.* Trans. A. Barker. In *Greek Musical Writings II.* Cambridge: Cambridge University Press, 1989.

———. *Die Harmonielehre des Klaudios Ptolemaios.* Ed. I. Düring. Göteborgs Högskolas Arsskrift 36 (1930).

———. *Opera astronomica minora.* Ed. J. L. Heiberg. Leipzig: Teubner, 1907.

———. *L'optique de Claude Ptolémée dans la version latine d'après l'arabe de l'émir Eugène de Sicile.* Ed. A. Lejeune. Louvain: Université de Louvain. Recueil de travaux d'histoire et de philologie, 4e série, fasc. 8 (1956).

———. *Ptolemy's Almagest.* Trans. G. J. Toomer. New York: Springer-Verlag, 1984.

———. *Syntaxis mathematica.* 2 vols. Ed. J. L. Heiberg. Leipzig: Teubner, 1898-1903.

———. *Tetrabiblos.* Ed. F. Boll and A. Boer. Leipzig: Teubner, 1957.

———. *Tetrabiblos.* Ed. and trans. F. E. Robbins. Cambridge: Harvard University Press, 1980. Loeb.

Sextus Empiricus. *Opera.* Ed. Hermann Mutschmann, rev. J. Mau. 4 vols. Leipzig: Teubner, 1968.

Simplicius. *In Aristotelis De caelo commentaria.* Ed. J. L. Heiberg. *Commentaria in Aristotelem Graeca.* VII. Berlin: Georg Reimer, 1894.

Stobaeus, Ioannes. *Anthologii qui inscribi solent eclogae physicae et ethicae.* Ed. C. Wachsmuth. 5 vols. Berlin: Weidmann, 1884–1912.

Stoicorum Veterum Fragmenta. Ed. J. von Arnim. 4 vols. Leipzig: Teubner, 1905–1924.

Strabo. *Geography.* Ed. and trans. H. L. Jones. 8 vols. New York: G. P. Putnam's Sons, 1917. Loeb.

Suidae lexicon. Ed. A. Adler. 5 vols. Leipzig: Teubner, 1928–38.

Theon of Alexandria. *Commentaires de Pappus et de Théon d'Alexandrie sur l'Almageste: Tome II. Théon d'Alexandrie. Commentaire sur les livres 1 et 2 de l'Almageste.* Ed. A. Rome. Studi e Testi 72. Città del Vaticano: Biblioteca Apostolica Vatican, 1936.

Theon of Smyrna. *Philosophi Platonici Expositio rerum mathematicarum ad legendum platonem utilium.* Ed. E. Hiller. Leipzig: Teubner, 1878.

Vitruvius. *Vitruvius On Architecture.* Ed. F. Granger. 2 vols. Cambridge: Harvard University Press, 1931. Loeb.

II. Writings by Modern Authors

Asmis, Elizabeth. *Epicurus' Scientific Method.* Ithaca, New York: Cornell University Press, 1984.

Barnes, Jonathan. *The Presocratic Philosophers.* London: Routledge and Kegan Paul, 1982, revision of 1979 ed.

Boll, Franz. "Studien über Claudius Ptolemäus: Ein Beitrag zur Geschichte der griechischen Philosophie und Astrologie." *Jahrbücher für classische Philologie,* supplement 21 (1894): 49–244.

Bouché-Leclercq, A. *L'Astrologie Grecque.* Paris: Leroux, 1899.
Bréhier, Émile. *The Hellenistic and Roman Age.* Trans. Wade Baskin. Chicago: University of Chicago Press, 1965.
Burkert, Walter. *Lore and Science in Ancient Pythagoreanism.* Trans. Edwin L. Minar. Cambridge: Harvard University Press, 1972.
Burnyeat, Myles, ed. *The Skeptical Tradition.* Berkeley: University of California Press, 1983.
Cherniss, Harold. *The Riddle of the Early Academy.* Berkeley: University of California Press, 1945.
Conroy, Donald Paul. *Epicurean Cosmology and Hellenistic Astronomical Arguments.* Ph.D. dissertation. Princeton University, 1976.
Copleston, Frederick. *A History of Philosophy: Volume I: Greece and Rome Part II.* Garden City, New York: Image Books, 1946, new revised ed., 1962.
Cornford, Francis MacDonald. *Plato's Cosmology: The Timaeus of Plato.* Indianapolis: Bobbs-Merrill, Library of Liberal Arts, rpt. 1975.
Cramer, Frederick Henry. *Astrology in Roman Law and Politics.* Philadelphia: American Philosophical Society, 1954.
Cumont, Franz. *Astrology and Religion among the Greeks and Romans.* New York: Dover, 1960, rpt. of 1912 ed.
———. *L'Egypte des Astrologues.* Brussels: Fondation Égyptologique, 1937.
———. *Oriental Religions in Roman Paganism.* Authorized translation. Chicago: Open Court, 1911.
Dillon, John. *The Middle Platonists 80 B.C. to A.D. 220.* Ithaca, New York: Cornell University Press, 1977.
Dodds, Erik Robertson. "Albinus." In *OCD.*
———. "Neopythagoreanism." In *OCD.*
Dörrie, Heinrich. "Der Platoniker Eudoros von Alexandreia," *Hermes: Zeitschrift für classische Philologie* 79 (1944): 25–30.
Edelstein, Ludwig. "The Philosophical System of Posidonius," *American Journal of Philology* 57 (1936): 286–325.
Ferguson, John. *The Religions of the Roman Empire.* Ithaca, New York: Cornell University Press, 1970.
Frazer, P. M. *Ptolemaic Alexandria.* 3 vols. Oxford: Clarendon Press, 1972.

Freudenthal, Jakob. "Albinus." In *RE*.
———. "Attikos." In *RE*.
Furley, David J. "Lucretius and the Stoics." *Bulletin of the Institute for Classical Studies of the University of London* 13 (1966): 13–33.
Gercke, Alfred. "Aspasius." In *RE*.
Grayeff, Felix. *Aristotle and His School: An Inquiry into the History of the Peripatos with a Commentary on Metaphysics* Z, H, Λ, *and* Θ. London: Duckworth, 1974.
Guthrie, W. K. C. *A History of Greek Philosophy*. 2 vols. Cambridge: Cambridge University Press, 1965, rpt. 1978.
Hahm, David E. *The Origins of Stoic Cosmology*. [Columbus]: Ohio State University Press, 1977.
Heath, Thomas. *Aristarchus of Samos: The Ancient Copernicus*. New York: Dover, 1981, rpt. of 1913 ed.
———. *Mathematics in Aristotle*. Oxford: Clarendon Press, 1949.
Helden, Albert van. *Measuring the Universe: Cosmic Dimensions from Aristarchus to Halley*. Chicago: University of Chicago Press, 1985.
Jaeger, Werner. *Aristotle; Fundamentals of the History of his Development*. Oxford: Clarendon Press, 1934.
Kirk, G. S., J. E. Raven, and M. Schofield. *The Presocratic Philosophers*. Cambridge: Cambridge University Press, 1983. 2nd ed.
Klein, Jacob. *A Commentary on Plato's Meno*. Chapel Hill: The University of North Carolina Press, 1965.
Knaack, M. "Eratosthenes von Kyrene (4)." In *RE*.
Kunitzsch, Paul. *Der Almagest: Die Syntaxis Mathematica des Claudius Ptolemäus in arabisch-lateinischer Überlieferung*. Wiesbaden: Otto Harrassowitz, 1974.
Lammert, Friedrich. "Eine neue Quelle für die Philosophie der mittleren Stoa I." *Wiener Studien: Zeitschrift für klassische Philologie* 41 (1919): 113–21.
———. "Eine neue Quelle für die Philosophie der mittleren Stoa II." *Wiener Studien: Zeitschrift für klassische Philologie* 42 (1920–21): 34–46.
———. "Hellenistische Medizin bei Ptolemaios und Nemesios. Ein Beitrag zur Geschichte der christlichen Anthropologie." *Philologus: Zeitschrift für das klassische Althertum* 94 (1941): 125–41.
———. "Kritische Untersuchung zu Ptolemaios. Περὶ

κριτηρίου καὶ ἡγεμονικοῦ." *Hermes: Zeitschrift für klassische Philologie* 72 (1937): 450–65.
———. "Ptolemais Περὶ κριτηρίου καὶ ἡγεμονικοῦ. und die Stoa." *Wiener Studien: Zeitschrift für klassische Philologie* 39 (1917): 249–258.
———. "Ptolemaios. 66. Klaudios Ptolemaios, der Astronom und Geograph." In *RE*.
———. "Zur Erkenntnislehre der späteren Stoa." *Hermes: Zeitschrift für klassische Philologie* 57 (1922): 171–88.
Long, A. A. "Ptolemy on the Criterion: An Epistemology for the Practising Scientist," in John Dillon and A. A. Long, *The Question of Eclecticism: Studies in Later Greek Philosophy*. Berkeley: University of California Press, 1989. Also in Pamela Huby and Gordon Neal, *The Criterion of Truth: Essays Written in Honour of George Kerferd*. Liverpool: Liverpool University Press, 1989.
Lloyd, G. E. R. *The Revolutions of Wisdom: Studies in the Claims and Practice of Ancient Greek Science*. Berkeley: University of California Press, 1987.
MacMullen, Ramsey. *Enemies of the Roman Order*. Cambridge: Harvard University Press, 1966.
Martini, Edgar. "Eudoros von Alexandrien (10)." In *RE*.
Mathewson, R. "Aristotle and Anaxagoras: an examination of F. M. Cornford's interpretation." *Classical Quarterly* (1958): 67–81.
Merlan, Philip. "Greek Philosophy from Plato to Plotinus," in *The Cambridge History of Later Greek Philosophy*. Ed. A. H. Armstrong. Cambridge: Cambridge University Press, 1967, pp. 14–136.
Mogenet, J. "L'Introduction à l'Almageste." Brussels: Palais des Académies, 1956. Académie royale de Belgique. Classe des lettres et des sciences morales et politiques. *Mémoires* 51, 2: 1–51.
Moraux, Paul. *Les listes anciennes des ouvrages d'Aristote*. Louvain: Éditions Universitaires de Louvain, 1951.
Mugler, Charles. *Dictionnaire historique de la terminologie géométrique des Grecs*. 2 vols. *Études et Commentaires* 28 and 29. Paris: Éditions Gauthier-Villars, 1958.
Neugebauer, O. *The Exact Sciences in Antiquity*. 2nd ed. New York: Dover, 1969.
———. *A History of Ancient Mathematical Astronomy*. 3

vols. Berlin, Heidelberg, New York: Springer-Verlag, 1975.

Owen, G. E. L. "The Platonism of Aristotle," *Proceedings of the British Academy* 50 (1965): 25–50. Reprinted in *Studies in the Philosophy of Thought and Action*. Ed. P. R. Strawson. London: Oxford University Press, 1968, and also in *Articles on Aristotle*. Ed. Jonathan Barnes, Malcolm Schofield, and Richard Sorabji. London: Duckworth, 1975.

Oxford Classical Dictionary. Ed. N. G. L. Hammond and H. H. Scullard. 2nd ed. Oxford: Clarendon Press, 1979, rpt. of 1970 ed.

Pedersen, Olaf. *A Survey of the Almagest*. [Odense]: Odense University Press, 1974.

Paulys Realencyclopädie der classischen Altertumswissenschaft. Stuttgart: Alfred Druckenmüller Verlag, 1893–1980.

Rist, John M. "Categories and their Uses," in A. A. Long, *Problems in Stoicism*. London: University of London, The Athlone Press, 1971.

Ross, William David. "Theon of Smyrna." In *OCD*.

Sambursky, S. *The Physical World of Late Antiquity*. Princeton: Princeton University Press, 1987, reprint of 1962 ed.

Schmekel, August. *Die Philosophie der mittleren Stoa in ihrem geschichtlichen Zusammenhange*. Berlin: Weidmann, 1892.

Solmsen, Friedrich. "Eratosthenes as Platonist and Poet," *Transactions and Proceedings of the American Philological Association* 73 (1942): 192–213.

———. *Plato's Theology*. Ithaca: Cornell University Press, 1942.

Swerdlow, N. M. *Ptolemy's Theory of the Distances and Sizes of the Planets: A Study of the Scientific Foundations of Medieval Cosmology*. Ph.D. dissertation, Yale University, 1968.

Tarán, Leonardo. *Academica: Plato, Philip of Opus, and the Pseudo-Platonic "Epinomis"*. Philadelphia: American Philosophical Society, 1975.

Taylor, A.E. "Plato and the Authorship of the 'Epinomis,'" *Proceedings of the British Academy* 15 (1929): 235–317.

Toomer, G. J. "The Mathematician Zenodorus." *Greek, Roman and Byzantine Studies* 13 (1972): 177–192.

———. "Ptolemy." *Dictionary of Scientific Biography* 11. New York: Charles Scribner's Sons, 1975.

Vlastos, Gregory. *Plato's Universe*. Seattle: University of Washington Press, 1975.

Von Fritz, K. "Theon aus Smyrna (14)." In *RE*.

Whittaker, John. "Platonic Philosophy in the Early Centuries of the Empire." *Aufstieg und Niedergang der Römischen Welt* (1987) Teil II, Band 36. 1: 81–123.

Witt, Reginald Eldred. *Albinus and the History of Middle Platonism*. Cambridge: Cambridge University Press, 1937.

Zeller, Eduard. *Aristotle and the Earlier Peripatetics*. Trans. B. F. C. Costelloe and J. H. Muirhead. 2 vols. New York: Russell and Russell, 1962, rpt. of 1897 ed.

———. *A History of Eclecticism in Greek Philosophy*. Trans. S. F. Alleyne. London: Longmans, Green and Co., 1883.

———. *The Stoics, Epicureans, and Skeptics*. Trans. Oswald J. Reichel. New York: Russell and Russell, 1962.

Ziegler, Konrat. "Ptolemaios. 66. Klaudios Ptolemaios, der Astronom und Geograph." In *RE*.

Index

Academy, 11–12, 144, 150
aether (*see also* elements, material; homoeomerous), 52–53, 56–58, 60, 133
  divinity of, 141–42, 145
Aetius, 145
air (*see also* elements, material)
  and arguments against motion of the Earth, 98–99
Albinus, 15
Almagest. *See* Ptolemy *Syntaxis*
Anaxagoras of Clazomenae, 54
Andronicus of Rhodes, 14
Antiochus of Ascalon, 11
Apollonius, 55
appearances. *See* observations
Arcesilaus of Pitane, 12
Aristotelian-Ptolemaic universe
  imprecision of the term, 2
Aristotle, *passim*
  aether, 141–45

celestial bodies, animate, 121; divine, 141–43; order of, 105–7
celestial-terrestrial distinction, blurring of, 123–24
direction terminology, 91–96
Earth, center of universe, 74–78, 97–98
Earth, immobility of, 84–100 (*passim*)
Earth, size of, 79–81
Earth, spherical shape of, 64–67
heavens, motion of, 59–60, 103, 115, 120–24
hypotheses, 41–44
influence on Ptolemy, *passim*, 9, 13–17, 19–28, 34, 36–37, 59–60, 64, 78–79, 84, 88–90, 96, 99–100, 119–25
Metaphysics, 14, 15, 20–24, 30, 115
Meteorology, 54

Nicomachean Ethics,
   19–21, 33–34
On the Cosmos,
   (pseudo-Aristotle),
   143–45
On the Heavens, 15,
   46–124 (*passim*), 145
On the Soul, 53
philosophy, division of,
   19–24
Physics, 24, 27–28, 51
physics, criticisms of,
   76–78; by Ptolemy,
   112–17
place, natural, 74–78
Posterior Analytics, 42
Aspasius, 15
astrology, 129–33

Boll, F., 9, 15, 21, 24

celestial bodies (*see also*
      aether)
   animate, 117–22
   distances of, 81–83,
      108–11, 131–32
   divine, 5–6, 53, 126,
      135–50
   harmonics, relation to,
      127–28
   motion of, 45–60, 100–103,
      111–25, 143–45, 148–49
   order of, 105–11, 131–32
   terrestrial region, relation
      to, 111, 122–25, 129–30
Cicero, 11, 15, 107
cosmos
   meaning of, 138–40, 143

Demiurge, 102, 147–48
Didaskalikos, 34, 151–52
Dillon, J., 8
Diogenes Laertius, 12, 141
Diotima, 31–33
directionality in the universe,
   91–97, 134
divine/divinity, 22, 25–27
   celestial bodies as, 5–6, 53,
      56, 126, 135–50
   imitation of, 33–34, 36–37,
      150–53
   and mathematics, 29–31

Earth
   center of universe, 71–79
   no local motion, 84–100
   size of, 79–84
   spherical shape of, 60–71
eclecticism, 9–13, 137
eclipse, 70–71, 108–9
Edelstein, L., 16
elements, material
   relation between celestial
      and terrestrial, 111,
      122–25, 129–30, 134
epicycles, 114
Epicurean philosophy, 76, 78,
   145
Epicurus, 37, 145–46
equatoria, 112
ethics, 5–6, 19–20, 30–33,
   141
   and astronomy, 36, 135–38,
      146–53
Euclid, 55, 83
   Elements, 42–43
Euctemon, 125
Eudorus of Alexandria, 15

fire (*see also* elements,
      material)
   celestial bodies composed
      of, in *Timaeus,* 148

Galen, 121
Geminus, 55, 83
geography, 68–71
gods (*see also*
      divine/divinity), 140–43,
      147–49
Goldstein, B. R., 119–20
Guthrie, W. K. C., 65–66, 139

harmonics, 125–28, 149
harmony, 143–45, 148–49
Heath, T., 28

Timaeus, 15, 92–96, 100, 102–3, 107, 141, 147–49
Pliny, 71, 107
Plutarch, 11, 107
  On the Face of the Moon, 77–78
Posidonius, 11, 15, 16
Prime Mover. *See* Unmoved Mover
Proclus, 42–43, 55–56
Ptolemy *passim*
  biography, 7
  ethics, 36, 135–38, 146, 151–53
  Geography, 69
  Harmonics, 4–5, 125–28
  influences, 8–17; Aristotle, *passim,* 9, 13–17, 19–28, 34, 36–37, 59–60, 64, 78–79, 84, 88–90, 96, 99–100, 119–25; Plato, 15, 31–34, 94–96, 121, 150–53
  On the Criteria, 9
  Phases of the Fixed Stars, 4, 125
  philosophy, 19–37
  Planetary Hypotheses, 4, 105, 107–23, 132–33
  Syntaxis, passim, 19–107
  Tetrabiblos, 4, 9, 15, 69, 105, 129–33, 136–37
Pyrrho of Elis, 12
Pythagoras, 121, 127
Pythagoreanism, 121, 141, 145, 151–52

religion, 10–13, 36–37, 137–38, 147–51
Rist, J. M., 16

Sambursky, S., 121

sciences, division of, 19–31
Seneca, 11
Sextus Empiricus, 12
Simplicius, 48–49
Skepticism, 12
Socrates, 31, 33
Soul, 102–3, 121
  human soul, 31, 126–28, 137, 147–50
Stobaeus, 76
Stoic philosophy, 9, 11–12, 15–17, 35, 76–78, 119, 129, 146
Strabo, 68–70
Suda, 7

terrestrial region, 23, 25, 123–25, 129–33
theology, 21–22, 25–30, 135–36
Theon of Alexandria, 54, 58
Theon of Smyrna, 15, 151–52
Theophrastus, 11, 129
  On Weather Signs, 125
Toomer, G. J., 2, 40–41, 44, 70–71, 119

Unmoved Mover, 113, 115–16, 120, 122

Vitruvius, 107
Vlastos, G., 138–39
void, 119

Xenocrates, 150

Zeller, E., 10
Zeno of Citium, 11, 76–77
Zenodorus, 52, 54–55, 57–58
  On Figure of Equal Boundary, 52
Zeus, 140, 145

heavens (*see also* aether, cosmos, celestial bodies)
    Earth has ratio of a point to, 79–84
    Earth in middle of, 71–79
    motions of, 45–60, 100–103, 111–25, 143–45, 148–49
Helden, van A., 2
Heraclides of Pontus, 98
Hesiodic poems
    Theogony, 140
    Works and Days, 125, 146
Hesychius of Miletus, 7
Hipparchus, 68, 70, 82
Homeric poems, 138–40
homoeomerous, 52–58
hypothesis, 39–45, 79, 125–26

immortality
    and divinity, 140, 142
    as goal, 31–33, 37, 149
infinity, 45–48
instruments, 49, 81, 112, 118, 125–26

Klein, J., 41–42
klima, 67–69, 81

Lacydes of Cyrene, 12
Lammert, 9, 15
Lamprias, 77–78
Lloyd, G. E. R., 5
Lucretius, 37, 76, 145

Marcus Aurelius, 7
mathematics, 19–31, 36–37, 135–36, 150–53
Meton, 125
motion (*see also* celestial bodies; Earth), 27–28
    natural motion/natural place, 65–66, 74–78, 85–88, 97–98, 122–23
music, 125–28, 134, 149, 151

Neoplatonism, 151–52

Neopythagoreanism, 11–12
Neugebauer, O., 2, 7, 9, 66–68, 70–71, 121
Nicomachus
    Introduction to Arithmetic, 151–52

observations, *passim,* 46–103
    relation to hypotheses, 40, 44–45, 96, 112

Odysseus, 138
Olympiodorus, 7

Panaetius, 11, 15, 129
Pedersen, O., 2, 15, 66, 120–21, 123
phenomena. *See* observations
Philo of Alexandria, 11
Philolaus, 141, 145
philosophy
    division of, 19–31
    goal of, 33–37, 149–53
    practical philosophy, 19–20, 31, 35, 36
    theoretical, 19–36
physics. *See* philosophy, division of
place, natural. *See* motion, natural
planets. *See* celestial bodies
Plato, 5, 11–12, 15–16, 36–37
    Epinomis (pseudo-Plato), 143–44, 150
    directionality, 91–96
    ethical motivation to study astronomy, 147–52
    hypotheses, 41–44
    influence on Ptolemy, 15, 31–34, 94–96, 121, 150–53
    Laws, 149
    Meno, 41
    planets, animate, 121
    Republic, 32–33, 41, 149
    Symposium, 31, 33
    Theaetetus, 33